ATLAS *of* CETACEAN *distribution in north-west European waters*

James B. Reid, Peter G.H. Evans and Simon P. Northridge

2003

Species texts written by
Jim Boran, Kees Camphuysen, Peter Evans, Mardik Leopold, Simon Northridge, Jim Reid, Mark Tasker and Caroline Weir

Reid, J B, Evans, P G H, and Northridge, S P (2003)
Atlas of Cetacean distribution in north-west European waters.
Joint Nature Conservation Committee, Peterborough.

CONTENTS

INTRODUCTION

Cetaceans sustain a depth of fascination that is almost without parallel in nature. Along with primates, they are among the most intelligent of all mammals; but while this might account for the allure of primates, the natural environment of cetaceans also adds to the interest of this order of animals. Encounters with cetaceans are difficult to contrive because they inhabit a world that is largely unseen. Despite increased attention in recent years, the oceans remain places of mystery; the ecological processes that underlie observable patterns in marine communities are largely unknown and seem almost arcane.

Compared with many terrestrial mammals, little is known of cetacean natural history. Indeed, knowledge of some species comes from only a handful of dead specimens, and new species are still being either recognised or discovered relatively frequently. Again, this is not surprising for animals that usually only briefly break the surface of the sea..

That cetaceans are difficult to observe, however, should not disguise the fact that populations of some species are very large. This is true not only in seas traditionally thought of as high in cetacean abundance but also of the Atlantic and associated seas around Europe. For example, in north-west European and adjacent waters there may be more than 100,000 minke whales present at certain times of the year (Schweder *et al.* 1997). The seas of north-west Europe also host a rich variety of cetacean species. More than 20 species may be seen here regularly throughout the year, about half as many again as occur regularly in the south-west Atlantic at similar latitudes (White *et al.* 2002). This represents a diversity of form and function that demands effective conservation.

The proper conservation of cetaceans depends on knowledge of many aspects of their population ecology. Ideally, information on population size, structure and seasonal distribution, as well as data on mortality, breeding productivity, and emigration and immigration rates should be available. For the reasons outlined above such information for most cetaceans is largely non-existent.

However, cetacean populations in some parts of the world are rather better known than in others. In north-west European waters, for example, there are many data available that may aid in the conservation of populations here. In particular, the increasing number of at-sea surveys adds to knowledge of the distribution, and in some cases the abundance, of our animals. Perhaps the best example here is the SCANS survey, which investigated the distribution and estimated the population size of harbour porpoises in the North Sea and adjacent waters in July 1994 (Hammond *et al.* 1995, 2002). Such large-scale surveys aid in placing subsequent mortality events, chronic or otherwise, in the proper population and geographical contexts. For example, the estimated magnitude of fishery bycatches of porpoises in the North Sea has allowed an assessment of the sustainability of current bycatch levels to be made (Vinther 1999; Hammond *et al.* 2002). Identification of patterns of cetacean distribution and abundance is an important early goal of research that aims to underpin conservation measures, as well as being one that can be relatively easily achieved. In conjunction with other information, knowledge of cetacean dispersion also aids in the investigation of the ecological determinants of dispersion, of the biological and physical processes that might generate dispersion patterns, and consequently of the habitat requirements of the various species. If time series data are also available then any detected distributional shifts, or changes in population size, of these top predators may be indicative of more far-reaching changes, or even disruption, to ecosystem processes.

It is perhaps as a natural heritage resource in their own right, however, that cetaceans are increasingly the focus of conservation research. The array of potential threats to cetacean populations has never been greater. Anthropogenic effects in the form of oil and chemical pollution, disturbance, noise pollution, habitat degradation and even deliberate persecution persist, although impacts on populations remain either minimal or obscure. While hunting currently poses little or no threat, notwithstanding the fact that the minke and pilot whales are still exploited by Norway and the Faroe Islands in the north-east Atlantic, interactions with fisheries do result in detectable changes at the population level. It is now widely accepted that fisheries have played a major role in the dramatic decline of porpoises in the Baltic in recent decades (ASCOBANS 2002), and there is also serious cause for concern that current or recent bycatch levels of porpoises in the North Sea and Celtic Sea may be unsustainable (Harwood *et al.* 1999; Tregenza *et al.* 1997).

For whatever reason, populations of some species in the north-east Atlantic have been extirpated from localised areas (e.g. the harbour porpoise in the eastern Channel) or even from much larger areas (e.g. the grey whale in the Atlantic).

The increasing diversity of possible threats to cetacean populations has seen a concomitant rise in the number of government agencies and non-governmental organisations (NGOs) devoted to their study. The former have instigated investigations on national and international scales, while NGOs have focused mostly on more restricted areas. In the UK, the principal and longest established NGO that carries out conservation research into cetaceans in British and Irish waters is the Sea Watch Foundation, which has been doing so in collaboration with other, smaller NGOs since 1973. The UK Government's advisor on nature conservation, the Joint Nature Conservation Committee (JNCC, formerly the Nature Conservancy Council), has also, since 1979, been carrying out research on the distribution of cetaceans from a much wider sea area of the north-east Atlantic. In association with other European government bodies and NGOs (listed in the Acknowledgements), the JNCC acts as a focus for the collation of effort-related sightings data of cetaceans over the north-west European continental shelf. The Sea Mammal Research Unit, part of the UK Government's Natural Environment Research Council and now also of the University of St Andrews, Scotland, has also been pursuing collaborative research on cetaceans with international conterparts since its formation in 1978. These organisations, with the longest track records of cetacean research in Europe, and their collaborators have contributed cetacean sightings data to a co-operative venture, the Joint Cetacean Database (JCD), and it is the data from this resource that are depicted in this Atlas.

The JCD comprises the most comprehensive information on the distribution of cetaceans in north-west European waters and this Atlas contains the most complete quantitative description of cetacean dispersion in this region. The distributional data herein presented for many of the more commonly occurring species are, for the first time at this broad geographical scale, effort-related; the user of the Atlas is thus accorded a greater degree of interpretative scope than previously available. The Atlas and database therefore aim to function as a practical conservation tool; as a first step in an audit of the occurrence of cetaceans at this scale their use can, with caution (see Methods), enable the proper contexting of new information on cetacean distribution and abundance, and they may also inform efforts to identify areas that might be particularly important for cetaceans. Indeed the database has already fulfilled this latter function with respect to the possible identification of Special Areas of Conservation for harbour porpoise (Bravington *et al.* 2002; Evans and Wang 2003).

The JCD is the product of a very fertile collaboration between different organisations both within the UK and in Europe as a whole. In the longer term, the JCD has the potential for growth as the existing collaborators continue to collect and contribute data, and as new partners join the venture. Such development should widen the scope and utility of the database and may enable identification of seasonal patterns of cetacean dispersion as well as habitat associations; perhaps it will foster more process-related research.

This Atlas aims to provide an account and snapshot of the distribution of all 28 cetacean species that are known certainly to have occurred in the waters off north-west Europe in the last 25 years, but including also narwhal and melon-headed whale for which records are as recent only as the 1940s. It cannot function as a 'where to watch cetaceans' guide and the reader is advised to read carefully the Methods chapter in order to aid interpretation of the maps. Most of the book comprises chapters covering individual species. In the majority of these, a brief account of the natural history of the species is presented, including information on identification, behaviour and social organisation, diet, and habitat preferences, inasmuch as such information is known. There follows some details of the species' worldwide distribution and its status in the north Atlantic, and then a description of its occurrence in north-west Europe accompanied by a map depicting this.

The Methods chapter describes data collection methods, database establishment, brief details of the analytical methods that were applied to render the data from many sources compatible, and the important section on map interpretation. In other chapters, the nature of the marine environment of the study area is described and information is presented on the current legislative instruments aimed at protecting cetaceans.

THE MARINE ENVIRONMENT

The topography of the Atlantic Ocean floor is dominated by a central S-shaped ridge extending from the Arctic Ocean south as far as the Southern Ocean. This immense mountain range - the mid-Atlantic ridge, divides the Atlantic into two roughly parallel troughs, which are in turn sub-divided into smaller basins by transverse ridges that emerge to form islands such as Iceland, the Faroes, and the Azores. Much of the North Atlantic Ocean here reaches depths of 1,000-5,000 m. Further east, the continental shelf of north-west Europe gives rise to the islands of Britain, Ireland and their satellites. The seas off north-west Europe, including the North Sea, the Irish Sea and the English Channel, are much shallower and only in a few places exceed a depth of 200 m. On the northern edge of the Bay of Biscay, to the north-west of Ireland, and west of the Outer Hebrides, the continental shelf slopes very steeply from 200 m to 2,000 m or more over a distance of a few kilometres.

The Atlantic region

The current pattern of the North Atlantic is dominated by the great sweep of the North Atlantic Gyre, a roughly circular system of relatively warm waters driven by the North East Trade Winds between 10° and 30° N, and by westerlies between 40° and 60° N. The Gulf Stream (or North Atlantic Drift or Current as it becomes known further north) affects profoundly the climate of Europe. It has its origins near the Bahamas and sweeps across the Atlantic and extends far into the Barents Sea, the main branch passing to the north-west of Scotland. It has several offshoots; the Canaries Current turns southwards, another turns east towards southern Ireland, southern England and Brittany, the Irminger Current turns north towards east Greenland, and another current turns into the North Sea between Scotland and Norway. The speed of the main circulatory currents is usually about 10 km per day whereas the Gulf Stream is much faster and can exceed 125 km per day.

The two deep troughs that lie between Britain and Rockall and Britain and the Faroe Islands are important features in determining deep water circulation in the region (Ellett *et al.* 1983). Cold, sub-zero waters of the Arctic flow southwards at depths below 500 m in the Shetland-Faroe Channel but are deflected westwards by the Wyville Thomson Ridge, which rises to a depth of about 500 m). The northeastwards flowing North Atlantic drift interacts with this deep water in a complex and turbulent manner, particularly on each side on the Channel. This turbulence can force deeper water, richer in nutrients towards the surface where it will enhance productivity. Similar turbulence caused by the steep sides of the offshore banks and seamounts also may enhance primary productivity over distances of several hundreds of metres.

The Norwegian Rinne is a trough that lies just west of southern Norway and also influences oceanic circulation in the region. Some of the water moving northeastwards to the west and north of the British Isles turns southwards into the North Sea at several points. A relatively small flow enters via the Fair Isle channel between Orkney and Shetland, with larger flows southwestwards to the east of Shetland and along the western edge of the Rinne.

The open ocean is exposed to wind, and gales are frequent with big seas built up by the prevailing westerlies, resulting in the regular mixing of the surface layers and the dispersal of plankton and fish. Gelatinous animals and planktonic crustaceans dominate the fauna of the surface waters, whereas at *c.* 300-600 m deeper, various species of cephalopod may aggregate, forming prey patches for several species of cetacean such as sperm whale, the beaked whales, pilot whales and pelagic dolphins. Fish are less abundant in the deep ocean than over the shelf although they may form seasonal aggregations or concentrate over banks.

Along the shelf slope, the Atlantic Ocean water masses meet the less saline waters of the continental shelf, which receive freshwater inputs from rivers in Britain, Ireland, and continental Europe. Although there is relatively little mixing along this boundary, currents of slightly warmer water move northwards along this shelf edge from the Celtic and Irish Seas to Shetland carrying plankton, including fish eggs and larvae, from south to north. Turbulence brings deep water up the shelf slope to within 200 m of the surface, resulting in enhanced productivity of plankton and associated aggregations of cephalopods and fish such as blue whiting and mackerel. These in turn attract concentrations of pelagic seabirds and species of cetacean such as the larger baleen whales, pilot whales, killer whales, common, and Atlantic white-sided dolphins (Evans 1990).

Most of the continental shelf on the Atlantic side is exposed to the prevailing southwesterly winds, and saline oceanic water crosses the shelf edge between Malin Head in north-west Ireland and Barra Head in the Outer Hebrides, often intruding over other parts of the shelf in winter. Along the west coast of Ireland, the Irish Shelf Current flows northwards and then eastwards along the north coast. Waters to the west of this, governed by the North Atlantic Current, are separated by the Irish Shelf Front and flow north off the northern half and south off the southern half of Ireland (McMahon *et al.* 1995).

Frontal systems occur where two water masses of different densities meet; such density differences may be generated by temperature or salinity or both. The main front in the Atlantic region is the Irish Shelf Front that occurs to the south and west of Ireland (at *c.* 11° W) around the 150 m isobath and exists year-round (Huang *et al.* 1991). This front marks the boundary between water over the Irish shelf (often mixed vertically by the tide) and offshore North Atlantic water. The turbulence caused by the front may bring nutrients from deeper water to the surface where they may promote the growth of phytoplankton, especially of diatoms in spring, but also dinoflagellates where there is increased stratification.

These may in turn be fed on by swarms of zooplankton and associated with these, aggregations of fish, seabirds and cetaceans.

The Central region including the Irish Sea

The Irish Sea is shallow (less than 100 m deep in most places) and largely sheltered from the winds and currents of the North Atlantic although its relatively high salinity indicates the influence of oceanic water from the south. In the Irish Sea, the inshore Coastal Current carries water from St George's Channel northwards through the North Channel, mixing with water from the outer Clyde. Southerly winds can strengthen this current, increasing the northward transport of water from the Irish Sea into the Sea of Hebrides, with northerlies retarding the current. Strong tides flow through St George's Channel and the North Channel in and out of Liverpool Bay and the Solway Firth, leaving an area of almost permanently slack water off the Irish coast north of Dublin. At the boundary between the fast moving mixed water of the tidal stream and the stratified slack water, the Irish Sea Front forms between the south coast of the Isle of Man and the coast of County Dublin (Pingree and Griffiths 1978). Such tidal mixing fronts are often zones of high biological activity (Pingree *et al.* 1978), where plankton growth and activity can be much higher than in adjacent stratified and mixed zones, due to elevated nutrient levels. The Irish Sea Front exhibits little variability in either position or structure and is particularly well-developed in August (Simpson 1981; Huang *et al.* 1991).

Waters immediately to the north of the front at this time hold high concentrations of harbour porpoises.

Seasonal fronts occur at several other locations immediately west of Britain, including the Celtic Sea Front west of the Pembrokeshire Islands and the Islay Front between Islay and the coast of Northern Ireland (Pingree *et al.* 1978; Simpson 1981). The Islay Front persists through the winter due to stratification of water masses of different salinity (Hill and Simpson 1989). Similarly, where tides are only moderate, uneven bottom topography can have a considerable mixing effect, as for example in the Sea of Hebrides.

Also, eddies that occur downstream of headlands and islands (Pattiaratchi *et al.* 1986), and narrow channels that produce very strong local tides (Pingree and Mardell. 1986), can lead to mixing, which results in small-scale convergences, divergences and shear zones that occur in a tidal rhythm (Hamner and Haury 1977), favouring biological productivity and associated aggregations of fish, seabirds and cetaceans (Evans 1990; Webb *et al.* 1990).

The North Sea region and English Channel

The waters of the continental shelf are influenced not only by deep Atlantic water but also by the effects of land. Further east, such as in the eastern part of the English Channel and the Southern Bight of the North Sea, the effects of the Atlantic water diminish, although in winter there is a stronger inflow of warm, highly saline Atlantic water to the North Sea through the English Channel. This inflow, along with that mentioned earlier into the northern North Sea, together with the effects of the Earth's rotation, drive a cyclonic pattern of circulation in the North Sea (North Sea Task Force 1993).

Topographically, the North Sea comprises three parts: the Southern Bight (51-54° N) with water depths generally less than 40 m; the central North Sea (54-57° N) with water depths of 40-100 m (except for shallower areas on the Dogger Bank and coastal banks off western Denmark; and the northern North Sea (north of 57° N), which includes an area of shelf water 100-200 m deep and the Norwegian Rinne, with water depths from 200 m to more than 700 m in the Skagerrak between Denmark and Norway (Holligan *et al.* 1989).

In the North Sea, the strongest tidal currents occur in the Southern Bight, the German Bight, off the eastern coast of Scotland, and between Orkney and Shetland. As elsewhere, the combination of variations in water depth and in tidal currents leads to the development of distinct hydrographic regimes during the summer months when a seasonal thermocline extends over most of the central and northern North Sea in response to solar warming. The transitional or frontal zones are characterised by strong horizontal gradients in surface or bottom water temperatures. In contrast to the shelf waters, which are well mixed in the winter by tidal action and winds, the deeper waters of the Norwegian Rinne exhibit thermohaline stratification throughout the year.

Examples of frontal zones may be found around the Frisian Islands and Helgoland, east of Spurn Head and Flamborough Head in Humberside, in the outer Firths of Forth and Tay, east of Buchan in Grampian, and between Orkney and Shetland (Pingree and Griffiths 1978; Holligan *et al.* 1989).

Seasonal changes in surface temperature are most pronounced in the southern and eastern parts of the North Sea, where the water is relatively shallow and influenced by the more extreme continental climate. Seasonal variations in surface salinity are relatively small, the most significant being the decrease in salinity of the Norwegian coastal waters during summer as a result of relatively fresh water flowing out from the Baltic Sea. There can, however, be large variations in mean temperature and salinity distributions from year to year depending upon climate.

The eastern sector and coastal areas of the English Channel are shallow, with depths rarely exceeding 50 m. Depths are greater in the central zone and generally slope from east to west reaching 100 m along the western edge, although a trough to the north-west of the Channel Islands, the Hurd Deep, reaches a depth of more than 170 m. Currents flow eastwards, bringing more saline water from the Atlantic. A frontal system (termed the Ushant Front) develops in summer in the transitional zone between cooler, tidally mixed Channel water and the warmer stratified water of the Atlantic (Pingree *et al.* 1978).

SURFACE CURRENTS IN THE NORTH-EAST ATLANTIC

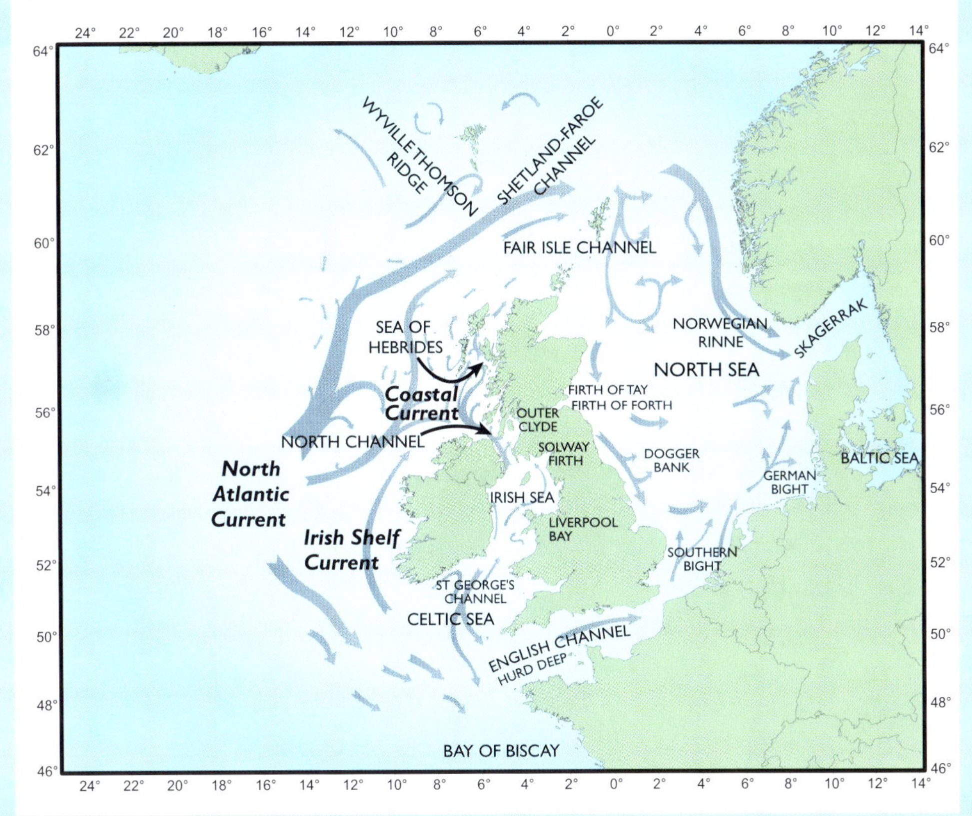

Figure 1

Schematic representation of major surface currents in the north-east Atlantic. The locations of selected habitat features are also indicated. (Adapted from Lee and Ramster 1981, Huang *et al.* 1991, Turrell *et al.* 1992, Hill 1993).

MAJOR FRONTS IN NORTH-WEST EUROPEAN WATERS

Figure 2

Locations of the major fronts in north-west European waters. The positions of the fronts indicated are approximate as there is a significant degree of temporal and spatial variation in their occurrence. (Adapted from Pingree and Griffiths 1978, Huang *et al.* 1991).

METHODS

Data Sources

Data on cetacean distribution and abundance presented in this Atlas came from three main sources.

The Seabirds at Sea Team (SAST) of the JNCC has been studying the distribution and abundance of seabirds and marine mammals over the (European) continental shelf since 1979. The data from at-sea surveys have been joined with similar data collected by sister organisations in other European countries (see Acknowledgements) to form the European Seabirds at Sea (ESAS) database. This database contains 1.7 million seabird records and over 13,000 cetacean records (of *c.* 49,000 individual animals) relating to the year-round dispersion of these taxa.

The UK Mammal Society Cetacean Group, subsequently forming the Sea Watch Foundation, has been collecting sightings data on marine mammals from UK and Irish waters since 1973, from opportunistic sightings and effort-related recording. Around 3,000 persons have contributed data over this period although most information comes from a much smaller number conducting regular watches either from land or on offshore surveys. The resultant database comprises 53,000 sightings records, of which 37,000 are opportunistic and the remaining 16,000, representing 84,000 individual animals, are associated with quantified survey effort. (Subsequent to the present work more of the 37,000 'opportunistic' records have been linked to quantified survey effort). Although all months of the year and all regions of the British Isles have been covered, effort has been greatest between April and September, within 30 km of the coast, and in certain regions (off north and west Scotland, the Irish Sea, and the Channel).

The Sea Mammal Research Unit co-ordinated the EC-funded SCANS survey (Small Cetacean Abundance in the North Sea) in June and July of 1994. The project involved intensive line transect sampling over a wide area from 62° N throughout the North Sea, Skagerrak and Kattegat, into the Western Baltic, and through the English Channel and into the Celtic Sea. The survey was and remains the largest synoptic abundance estimate for cetaceans in European waters, and covered about 20,000 km of trackline by boat and 7,000 km by air. Estimates of abundance were calculated for three cetacean species (Hammond *et al.* 2002).

Data from each of these three major sources were transformed to a common and simple format consisting of sightings records and effort records; this procedure is described further below. Only those sightings that could be associated with a measured amount of search effort were included in the subsequent analysis for the preparation of the maps in this Atlas. Overall, some 61,000 hours or 2,500 days of observation, and 31,000 sightings of 138,000 individual cetaceans are represented.

European Seabirds at Sea data

Surveys are conducted from moving platforms by several organisations in Europe, largely using the methods described by Tasker *et al.* (1984) and Webb and Durinck (1992). Many ships of opportunity were used for surveys, including research vessels, fishery protection vessels, seismic vessels and ferries; dedicated survey vessels were also used on occasion. Data were collected when the vessel was steaming, usually at speeds greater than 5 knots; its position, speed and course were recorded using a Global Positioning System (GPS). Cetaceans were also recorded from a small number of aerial surveys using similar, standardised methods (Pihl and Frikke 1992; see also Pollock *et al.* 2000).

Environmental data such as wind direction and force, sea state, swell height and visibility were recorded every 90 minutes, or more frequently if environmental conditions changed during an observation session.

Although surveys were designed primarily for the detection of seabirds, all cetacean sightings were also recorded, usually within 10 minute intervals. While some methodological differences existed between surveys carried out by different organisations, generally all individual animals observed within 90° of the ship's track line out to a perpendicular distance of 300 m were recorded within distance bands: A = 50 m, B = 51-100 m, C = 101-200 m, D = 201-300 m; animals were also recorded at distances greater than 300 m from the track line. During some surveys, including those conducted by the JNCC, observers used the naked eye to detect cetaceans, and binoculars to confirm species identification if necessary; however, during other surveys carried out by other organisations (mainly in the North Sea), binoculars were used to scan the sea within 2 km on both sides of the survey ship, and radial distance and angle to the animals were taken on first sighting. The species, numbers, age (if possible) and behaviour of the animal(s) were recorded during all surveys.

Base data allowed cetacean observations to be associated with specific locations, but within a 10 minute spread. Base position, speed and direction were recorded at the start and finish of each recording session, and also within sessions if the survey base changed speed or course. This allowed interpolation to define the position of the survey base, usually at 10 minute intervals, thereby allowing the association of species data with unique geographical locations.

Sea Watch data

From the early 1960s to the late 1980s, with the notable exception of certain land-based observation points (such as bird observatories), most sightings hosted by the Sea Watch database were opportunistic and not from dedicated watches. Subsequently, most sightings have been effort-related, usually from dedicated watches and made both from land and offshore.

Land-based watches were made mainly from headlands or islands of 1-3 hours duration, and, for a number of sites, conducted on a regular basis varying from frequent intervals (daily to weekly) between April and September to longer intervals (weekly to monthly) in winter (although some sites were not watched at all in winter).

During the course of the watches, visual scans with the naked eye were generally made to detect cetaceans, followed by binocular or telescope observations to confirm possible sightings and determine species identity, group size, presence of calves, and behaviour. Other information noted included behaviour (using standardised categories) and the presence of seabirds associated with the cetacean(s). Effort was recorded as the duration of the watch in minutes. Environmental data were usually collected at 10-30 minute intervals and included cloud cover, wind direction and force (using Beaufort scale), sea state, swell height, precipitation, and visibility. Offshore observations were made from a variety of platforms of opportunity including research vessels, fishery protection vessels, seismic vessels, whale watching vessels and ferries, and cruises dedicated to cetacean surveys. Until GPS was available, regular positions of the vessels were usually taken by DECCA; during the 1980s, these were increasingly replaced by GPS, which could also directly record the speed and course of the vessel, usually at 10-30 minute intervals. Vessel speed varied with the size and nature of the vessel but was rarely less than 5 knots.

Environmental data including cloud cover, wind direction and force (using Beaufort scale), sea state, swell height, precipitation, and visibility, were also recorded at 10-30 minute intervals. Dedicated watches for cetaceans generally involved naked eye scans (followed up by binocular checks to confirm species identity) to record all individuals observed within 90° of one side of the ship's track line for larger vessels (e.g. ferries), and 180° forward scans for smaller vessels; the latter accounts for most observations. All sightings out to the horizon were recorded, along with the time of day and vessel's position on first sighting, and, for some surveys, radial distance and angle to the animal(s). Other information recorded included group size, relative size of individuals (for detection of immatures and calves), behaviour (using standardised categories), direction of movement where this was obvious, and group sizes and species of associated seabirds.

SCANS data

The SCANS survey was conducted over a short time period (June and July 1994), and involved nine ships and two aircraft on dedicated sightings tracks. Aerial data were not included in the production of the present Atlas. Each ship had two independent sightings platforms with two or three observers on each platform. One set of observers used binoculars to spot cetaceans ahead of the ship, while on the other platform observers used the naked eye to detect cetaceans closer to the ship. Times, angles and radial distances were recorded for each cetacean sighting. The trackline course and vessel speed was recorded throughout each survey day, and 99% of all survey effort was conducted in sea states of 4 or less. The methods are described in further detail by Hammond *et al.* (2002).

The SCANS survey covered 16 sea-area blocks, and within each survey block zigzag cruise tracks were made to accord every point in the survey block a known non-zero probability of coverage. This survey design is intended to enable an unbiased estimate of abundance to be calculated regardless of the distribution of animals within each block. For the Atlas, only encounter rates were used, and both platforms on each ship were treated as independent vessels surveying close to one another at the same time (for harbour porpoises, fewer than 10% of sightings were duplicated by both platforms).

Data Treatment

As is clear from the above, data from each of the three major survey programmes are stored in different formats. The production of an Atlas in which all three data-sets are used therefore required a degree of data normalisation. Each of the three data-sets comprises records that may be described as 'sightings' records and 'effort' records. Sightings records provide information on the animals, including the species identity or other descriptor, the number of individuals and occasionally other details, including behaviour. The sightings record also provides a way of linking the sighting event with a 'leg' or unit of survey effort, where this is either an observation period of known duration or a distance travelled between a start point and an end point. Effort records include the start and end location, time and date of the observation period, details of the observer or observers, and certain key environmental parameters, including sea state.

In order to provide a standardised description of animal distribution, sightings rates are used to describe the perceived density of animals in particular areas. Usually, sightings rates are presented as encounters per km of survey track line, but because many data used in the Atlas were collected by stationary observers from vantage spots such as cliff-tops in timed intervals, the number of individuals sighted per unit time was deemed a more appropriate measure of sightings rate here.

The maps presented here depict the number of individuals of a particular species sighted per unit time of observation, resolved into 1/4 International Council for the Exploration of the Sea (ICES) rectangles (15' latitude x 30' longitude). The area of 1/4 ICES rectangles clearly varies with latitude, but is somewhat less than 1,000 sq km in the present study area. In presenting the data in this way, we make a number of important assumptions. We assume that species were identified correctly and that there is no bias in the estimated numbers of animals per group for each sighting. More importantly, we assume that there are no other sorts of biases associated with the effort legs.

One of the most obvious sources of potential bias arises from the sea state conditions that prevailed during each observation period. For many cetacean species the probability of visual detection decreases with increasing sea state; this is especially true for less conspicuous species such as the harbour porpoise and the minke whale. As sea states are generally higher in offshore areas and in winter, sightings rates in such areas and in the winter months are likely to be biased underestimates. In order to compensate for this bias, correction factors were used in the calculation of sightings rates for several of the species under consideration. Sightings rates were modelled as a function of several co-variates including sea state using general additive modelling (GAM; Bravington *et al.* 2001). Smoothed functions relating sea state to sightings rate were generated for eight species or species groups (see below; fin, sei and humpback whales were treated as the same species with respect to detectability), and these were then used to adjust survey effort within each sea state category by an appropriate correction factor. In effect, survey effort in higher sea states was down-weighted compared with effort in low sea states.

Correcting survey effort for differences in sea state was the only measure that we applied to account for possible biases in the estimation of sightings rates. Although other factors might also affect the probability of sighting a cetacean, including the number of observers present, the speed of the observation platform (which can be zero), the eye height of the observer and the observer's experience, they were not used to correct sightings rates; in effect, therefore, we assumed that they did not bias sightings rates by area. These are important assumptions that militate against over interpreting details on the maps that we have produced.

In order to assign all observations and the survey effort associated with them to their appropriate 1/4 ICES rectangles or cells, each observation was first allocated to the cell in which it occurred. For effort data where the start and end locations fell within a single cell, the observation duration was allocated to that cell. Where parts of more than one cell were surveyed during the course of an effort leg, the respective amounts of time (effort) spent in each of the cells were apportioned to the appropriate cells.

For each year and each month of the year, the total amount of time spent observing during each sea state category, and also the number of individuals of each species seen, was calculated for each cell of the study area. Effective effort was then computed for each cell (by area/time combination) by multiplying search effort (minutes) in each sea state category by the appropriate correction factor obtained from the general additive modelling, and summing the totals. Sightings rates were then computed as the number of individuals sighted per cell divided by effective search effort for that cell. These corrected sightings rates were then used in the maps. To reiterate, effective effort was calculated in this way for eight species or species groups and for these, therefore, effort was species-specific. For other species there were too few data to allow estimation of correction factors, so in these cases sightings rates are presented as numbers of animals observed per unit search time, uncorrected.

Map Interpretation

The maps of cetacean occurrence in the study area presented here are of three types: (a) those that depict the distribution, relative abundance and associated effort for those species for which sufficient data allowed estimation of correction factors (individuals/standardised hour). Maps of this type are presented for the following species: humpback whale, minke whale, sei whale, fin whale, short-beaked common dolphin, white-beaked dolphin, Atlantic white-sided dolphin, Risso's dolphin, long-finned pilot whale and harbour porpoise; (b) those that depict similar information for those species for which sufficient data did not exist to allow estimation of correction factors (individuals/hour). Maps of this type are depicted for sperm whale, northern bottlenose whale, selected beaked whales *Mesoplodon* spp., common bottlenose dolphin, striped dolphin and killer whale; and (c) sightings locations of rarely recorded species; two maps of this type are presented showing locations of northern right whale, blue whale and pygmy sperm whale, and of Cuvier's beaked whale, Sowerby's beaked whale and false killer whale. No maps are presented for species very rarely recorded or identified in north-west European waters, namely beluga, narwhal, Fraser's dolphin and melon-headed whale. The data used for each map span all years of data collection, from 1979 to 1997, so possible inter-annual shifts or differences in distribution are not reflected in the maps; rather, the maps represent an integrated picture or 'snapshot' over this 20 year period. Neither can these annual maps indicate any seasonal differences in distribution; maps depicting monthly distributions will be presented elsewhere (www.jncc.gov.uk).

On all maps except those for rare species, each individual grid cell, i.e. 1/4 ICES rectangle, is shaded to indicate the level of search effort achieved in that cell. The darker the shading, the more survey coverage achieved, while empty cells or blocks of cells indicate no survey coverage. The red dots that overlie this grid of rectangular cells indicate by their size the relative sightings rates for each species. Note that the minimum and maximum sightings rates on each map vary, as there are order of magnitude differences in the sightings rates between species. The maps cannot therefore be used to compare inter-specific differences in relative density. The red dots are intended solely to indicate relative density for a particular species, thereby enabling comparisons to be made only between areas for individual species.

The fact that sightings rates for some of the species presented here have been derived from search effort data that have been standardised for sea state differences also means that the grey effort shading differs among the species maps. Thus, for each species where sea state corrections have been applied to actual search effort, the corrected search effort is species-specific. Hence, search effort for the same cell in different species maps may differ (and frequently does).

It is worth reiterating that there are many potential biases associated with these data as presented. We have assumed that none of these biases is systematic, but in fact this is unlikely to be the case and only a detailed statistical analysis of the data will establish the nature and extent of any biases. The patchy nature of the observational coverage, however, should always be borne in mind when interpreting these maps. A chance sighting of a large school of animals during a relatively short observation period may lead to an apparently high relative density in a local area. However, the grey background shading indicating effort levels should enable the reader to filter the more extreme examples of such cases. Nevertheless, it would be inappropriate to infer too much about local densities of animals at the individual cell level. In our interpretation of these maps we have tried to avoid any such over-interpretation, and rely instead on general statements about relative animal densities at a regional level.

NORTHERN RIGHT WHALE

Eubalaena glacialis

The northern right whale is the rarest whale species that occurs in the north-east Atlantic. Reaching a length of around 16 m, it has a very robust form with a very large head comprising c. *30% of the total length, strongly arched upper jaws and strongly bowed lower jaws. The narrow rostrum has distinctive, light-coloured callosities that extend around the blowholes, on the chin and lower lips. The body is generally black, dark grey or dark brownish, sometimes mottled, with white patches on the throat and belly; the large, broad flippers are wholly black. The right whale lacks a dorsal fin and, on surfacing, may show a bushy V-shaped blow rising to 5 m.*

In recent times, the species has been seen either singly or in pairs or small groups of up to 12 individuals. Aggregations of up to 40 animals occur on the breeding and feeding grounds in the Bay of Fundy (Katona and Kraus 1999). Right whale social organisation is poorly known. Brownell and Ralls (1986) have speculated that several males competing for access to single females may actually engage in sperm competition. Mother-calf bonds are strong in the first 10 months of life and apparently unrelated individuals sometimes associate for several weeks at a time (Kraus and Evans, in press). Breaching, flipper-slapping, fluking up before diving, and lobtailing are all commonly seen at the surface.

The diet of right whales is exclusively zooplankton, mostly copepod crustaceans but also euphausiids, which are taken by open-mouth skim feeding (Mayo and Marx 1990). Areas of high concentrations of copepods *Calanus* spp. are favoured feeding grounds (Kenney *et al.* 1986). The species prefers coastal habitats - inshore waters, shallow bays, banks and estuaries. Mothers and calves have frequently been seen within 8-16 km of land, and sometimes within 1.6 km in very shallow water (Katona and Kraus 1999).

Global distribution and North Atlantic status

Northern right whales are restricted to the northern hemisphere between 20° and 65° N (i.e. mainly the temperate zone), predominantly in the north-west Atlantic. In the western North Atlantic, its range extends from Texas to the Gulf of St Lawrence and the coasts of Nova Scotia, Newfoundland and southern Greenland. Winter breeding appears to be concentrated in coastal Florida and Georgia; in spring, distribution is centred on Massachusetts Bay and in the Great South Channel, east of Cape Cod; and in summer and autumn, most animals congregate in the Bay of Fundy between Maine and Nova Scotia, and on the continental shelf 50 km south of Nova Scotia (Katona and Kraus 1999).

After centuries of over-exploitation, the North Atlantic population is a relict of its former size. Photo-identification studies suggest a population in the north-west Atlantic of around 325 individuals (a total of 380 individuals have been catalogued since 1980 - Kraus *et al.* in press). Numbers recorded on surveys in the north-east Atlantic are so small that it is not possible to estimate population size accurately; the few sightings would indicate no more than the low tens of individuals. It is possible that those animals recorded in the eastern North Atlantic have strayed from the western North Atlantic. For example, an individual seen off northern Norway in 1999 had previously been photographed in spring 1999 off the east coast of the USA, and was recorded there again in 2000. A survey of former breeding grounds in Cintra Bay on the coast of West Africa discovered no animals (Notarbartolo di Sciara, pers. comm.), but the species has been seen in other former breeding waters off the Canaries (E. Urquiola, pers. comm.).

Accorded protection from hunting in 1935, threats persist in the form of collisions with ships and entanglement in fishing gear which together are considered to be responsible for over 40% of all mortality of the species (Knowlton and Kraus, in press).

NW European distribution

In the eastern North Atlantic, the right whale once ranged from north-west Africa, the Azores and Mediterranean north to the Bay of Biscay, Channel and North Sea, western Ireland, the Western and Northern Isles of Scotland, Norway, the Faroe Islands, Iceland, and Svalbard. Since the 1920s, sightings have been sporadic, coming from the Canaries, Madeira, Spain, Portugal, the UK, Norway and Iceland (Brown 1986; Evans 1992, unpubl. data; E. Urquiola, pers. comm.).

In the 20th century, Scottish whalers took 94 animals off the Outer Hebrides and six off Shetland between 1906 and 1923 (Thompson 1928). None was obtained when whaling resumed in 1950-51.

Off western Ireland, 18 were caught from 1908 to 1910 but none in 1920 or 1922 (Fairley 1981). Since the inception of the British Museum (Natural History) stranding scheme in 1913, no right whales have stranded.

Most catches of the Scottish and Irish whale fisheries occurred in June, after which the species was thought to move to Scandinavian feeding areas later in the summer (Collett 1909; Thompson 1928). The few sightings in British and Irish waters have occurred between May and August (Evans 1992).

HUMPBACK WHALE

Megaptera novaeangliae

The humpback whale is a large baleen whale, adults growing to between 11.5 and 15 m in length. Many features distinguish it from other members of the family Balaenopteridae (rorquals). It has a slender, flattened head covered by fleshy tubercles with a rounded knob near the tip of the lower jaw. It has very long, often white flippers and, when it dives, its distinctly notched and irregularly edged tail flukes (with partial white undersides) are often thrown into the air. The head, front of the flippers and tail flukes are often infested with barnacles and whale lice. The dorsal fin is variable in size and shape from a small triangular knob to a larger sickle-shaped fin, situated two-thirds along the back.

Humpbacks are usually seen singly or in pairs; groups rarely exceed four or five animals unless in a feeding or breeding aggregation. In common with most baleen whales, the only stable social grouping appears to be the mother and calf; few other long-term associations have been observed (Clapham 1996). In the breeding grounds, social groups of up to several animals may form, with male escorts accompanying mother-calf pairs and behaving aggressively towards other males. The humpback is a very acrobatic species, exhibiting a variety of surface and underwater behaviour. They often breach, spyhop, lobtail, and tail slap.

The diet of humpback whales consists mainly of small schooling fish such as sandeels, herring, mackerel, capelin, pollack, cod, and anchovy, and large zooplankton, mainly euphausiids, but also invertebrates such as mysids and copepods (Winn and Reichley 1985). The type and amount of fish taken vary regionally, while euphausiids tend to be taken in greater quantities in polar regions.

A wide range of feeding behaviour has been recorded, including herding and disabling fish using the flipper, synchronised lunges, co-ordinated echelon feeding (where a group of animals swim side by side while feeding), and releasing a column or net of bubbles to entrap plankton (Winn and Reichley 1985). Humpback whales frequently feed in close proximity to other cetaceans, for example minke and fin whales and Atlantic white-sided dolphins. Seabirds such as black-legged kittiwakes, other gulls, and various auk species also associate frequently with feeding humpbacks.

Global distribution

The humpback whale occurs globally in tropical, temperate and polar seas of the northern and southern hemispheres. It favours waters over and along the edges of continental shelves, and around oceanic islands (Winn and Reichley 1985). In winter, humpbacks mate and give birth over shallow banks (commonly 15-60 m depth) in tropical waters (Whitehead and Moore 1982). In summer they tend to have a coastal distribution that is largely dependent on local prey availability (Winn and Reichley 1985).

North Atlantic status

Humpback populations, including that in the North Atlantic, have been severely depleted by over-exploitation. There are signs of recovery since they were accorded protection, at least in the north-west Atlantic.

Photo-identification studies indicate the north-west Atlantic population to number *c.* 10,600 individuals (95% CI: 9,300-12,100) (Smith *et al.* 1999), with an average annual rate of increase for the Gulf of Maine feeding stock of 6.5% (Barlow and Clapham 1997). In the north-east Atlantic, no population estimate exists, but Øien (1990) estimated the total population as much larger than previously thought, with main concentrations around Iceland and the Barents Sea. However, the Cape Verde Islands, once important as a whaling ground in the mid-1800s, currently appears to support few humpbacks (Reiner *et al.* 1996).

NW European distribution

The distribution map of effort-related sightings emphasises the rarity of the species, with isolated records almost exclusively in waters deeper than 200 m. However, in the 1990s, the number of casual sightings of humpbacks in UK waters increased from nine, between 1980-89, to 54 between 1990-99 (Evans 1996; Sea Watch, unpubl. data). Most records over the UK continental shelf have come from the Northern Isles (where up to three individuals have been seen annually since 1992), the northern Irish Sea and Firth of Clyde, and the southern Irish Sea, Celtic Sea and western Channel (where one or two individuals have been reported in most years since 1990) (Evans 1996; Sea Watch, unpubl. data). Most sightings have been made between May and September, which is when small numbers have also been seen off the continental shelf west and north of Scotland (Pollock *et al.* 2000).

HUMPBACK WHALE
Megaptera novaeangliae

Maximum sighting rate
0.38 animals / standard hour

Depth contours
200m
500m

Search effort as Standardised hours
0.1 to 10 hours
10 to 100 hours
More than 100 hours

HUMPBACK WHALE
Megaptera novaeangliae

Distribution data copyright
Joint Nature Conservation Committee
Sea Watch Foundation
Sea Mammal Research Unit

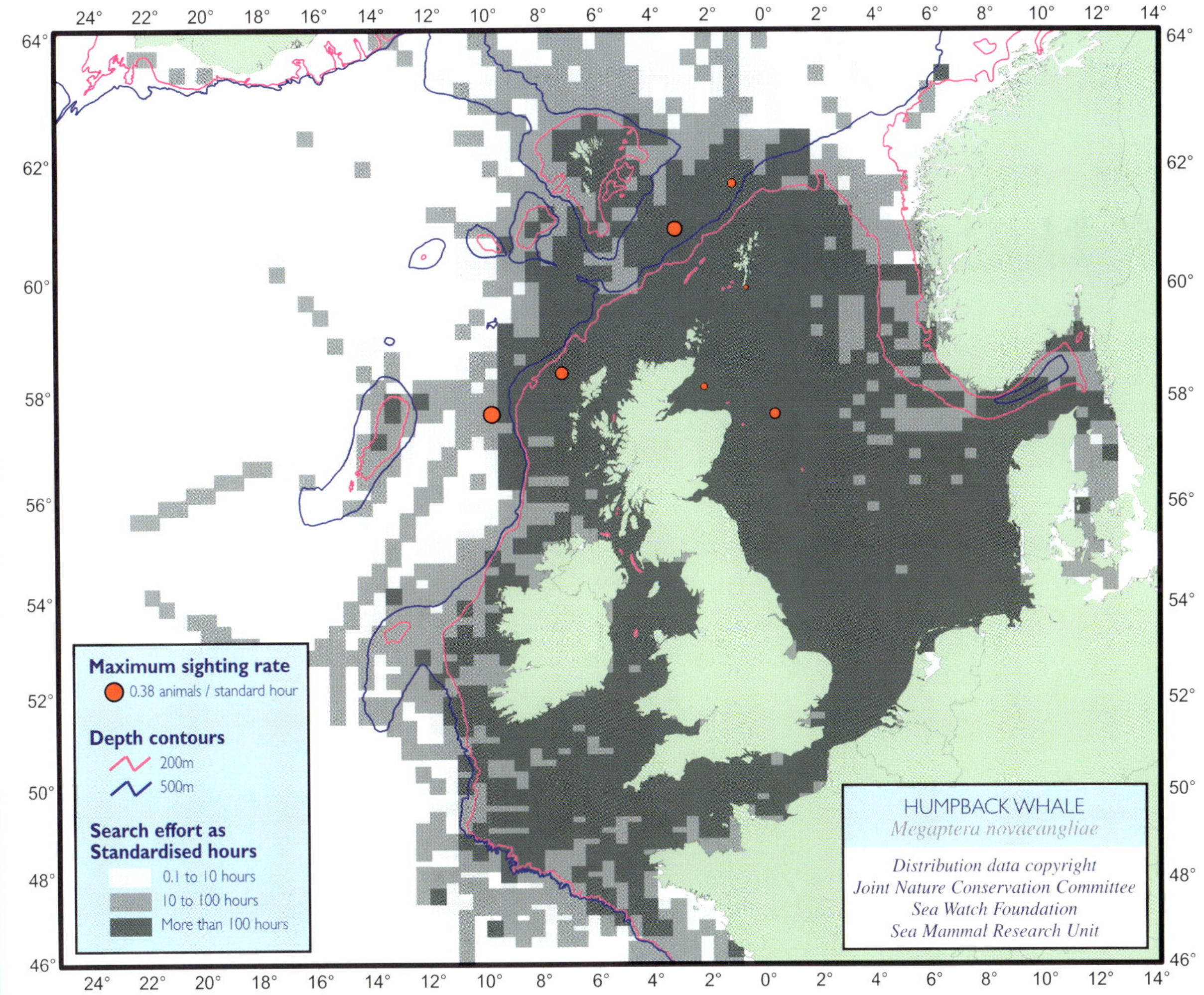

MINKE WHALE

Balaenoptera acutorostrata

The minke whale is the smallest of the rorquals, the balaenopterid family of baleen whales that includes also the blue, fin, sei and humpback whales. Growing to a length of 7-8.5 m, it is distinguished from other balaenopterids by a diagonal white band on its flipper, and a slender, pointed triangular head.

Minke whales are usually seen singly or in pairs although, when feeding, they sometimes aggregate into larger groups that can number 10-15 individuals. They frequently approach vessels and will occasionally both bow-ride and stern-ride. Breaching is not uncommon. When feeding, minke whales can often be seen in the vicinity of other cetacean species such as harbour porpoises and sometimes white-beaked and Atlantic white-sided dolphins, and humpback whales.

It is the most catholic feeder of all rorquals in the eastern North Atlantic, taking a wide variety of fish such as herring, cod, capelin, haddock, saithe, and sandeel, as well as euphausiids and pteropods (Haug *et al.* 1995; Nordøy *et al.* 1995). A variety of feeding methods are used, ranging from side- and lunge-feeding, where the animal swims swiftly up with open mouth to trap fish shoals at the sea surface, to the engulfing of prey with open mouth from behind. In Scottish waters, feeding minke whales in late summer are commonly associated with flocks of Manx shearwater, northern gannet, black-legged kittiwake, and various other gulls. Favoured feeding locations in summer include upwelling areas around headlands and small islands, often where strong currents flow (Evans 1990, 1991). There is some seasonal site fidelity (as revealed from photo-identification studies - e.g. Dorsey *et al.* 1990; Gill and Fairbairns 1995).

Global distribution

Minke whales are extensively distributed in northern hemisphere tropical, temperate and polar seas. In the North Atlantic, morphological differences suggest three distinct geographical populations of minke whales, from West Greenland, Icelandic and Norwegian waters (Christensen *et al.* 1990), and this has been supported by isozyme studies of genetic variation in minkes from the three localities (Daníelsdóttir *et al.* 1995). On the other hand, DNA studies of eastern North Atlantic minkes from Iceland and north Norway suggested mixing of these breeding stocks (Bakke *et al.* 1996).

North Atlantic status

In the North Atlantic, the species occurs from Baffin Bay in the west, and the Greenland and Barents Seas in the east, south to the Lesser Antilles in the west, and the Iberian Peninsula and Mediterranean in the east.

SCANS line transect surveys in 1994 estimated *c.* 8,500 (95% CI: 5,000-13,500) in the North Sea, Celtic Sea and Skagerrak (Hammond *et al.* 1995). Schweder *et al.* (1997) estimated the north-east Atlantic stock (seasonally inhabiting the North, Norwegian and Barents Seas) at 112,000 (95% CI: 91,000-137,000) in 1995.

NW European distribution

The minke whale is widely distributed along the Atlantic seaboard of Britain and

Ireland and also occurs throughout the northern and central North Sea as far south as the Yorkshire coast. It appears to be more abundant in the western part of the North Sea (but with a cluster of sightings in the centre of the North Sea between 56°30' and 58°30' N and 0-2° E). Although effort has been much greater nearer the east coast of Britain, the SCANS survey that spanned the entire North Sea in July 1994 also recorded much higher densities in the western sectors (Hammond *et al.* 1995). The species is rare in the southern half of the North Sea south of a line drawn from Humberside in eastern England to the north coast of Jutland in Denmark.

In Ireland, the minke whale is widely distributed, but most common along the south coast. It is more localised in the Irish Sea, and rare in the Channel, where it occurs almost exclusively in the western part; there is a concentration of sightings around the Brittany coast as well as on the northern edge of the Bay of Biscay.

The species occurs mainly on the continental shelf in water depths of 200 m or less. It can often be seen close to land, where it sometimes enters estuaries, bays or inlets. However, sightings have been reported far from shore between north-west Scotland and the Faroes in August, and over the Rockall bank in September, in waters of mainly 150-500 m depth.

Although the species occurs year-round on the north-west European continental shelf, most sightings have been made between May and September; during July to September, aggregations of feeding individuals can be observed, particularly near shore. There have been few records between October and April (mostly south of 50° N), although effort and viewing conditions are much poorer at this time.

MINKE WHALE
Balaenoptera acutorostrata

Maximum sighting rate
2.5 animals / standard hour

Depth contours
200m
500m

Search effort as Standardised hours
0.1 to 10 hours
10 to 100 hours
More than 100 hours

MINKE WHALE
Balaenoptera acutorostrata

Distribution data copyright
Joint Nature Conservation Committee
Sea Watch Foundation
Sea Mammal Research Unit

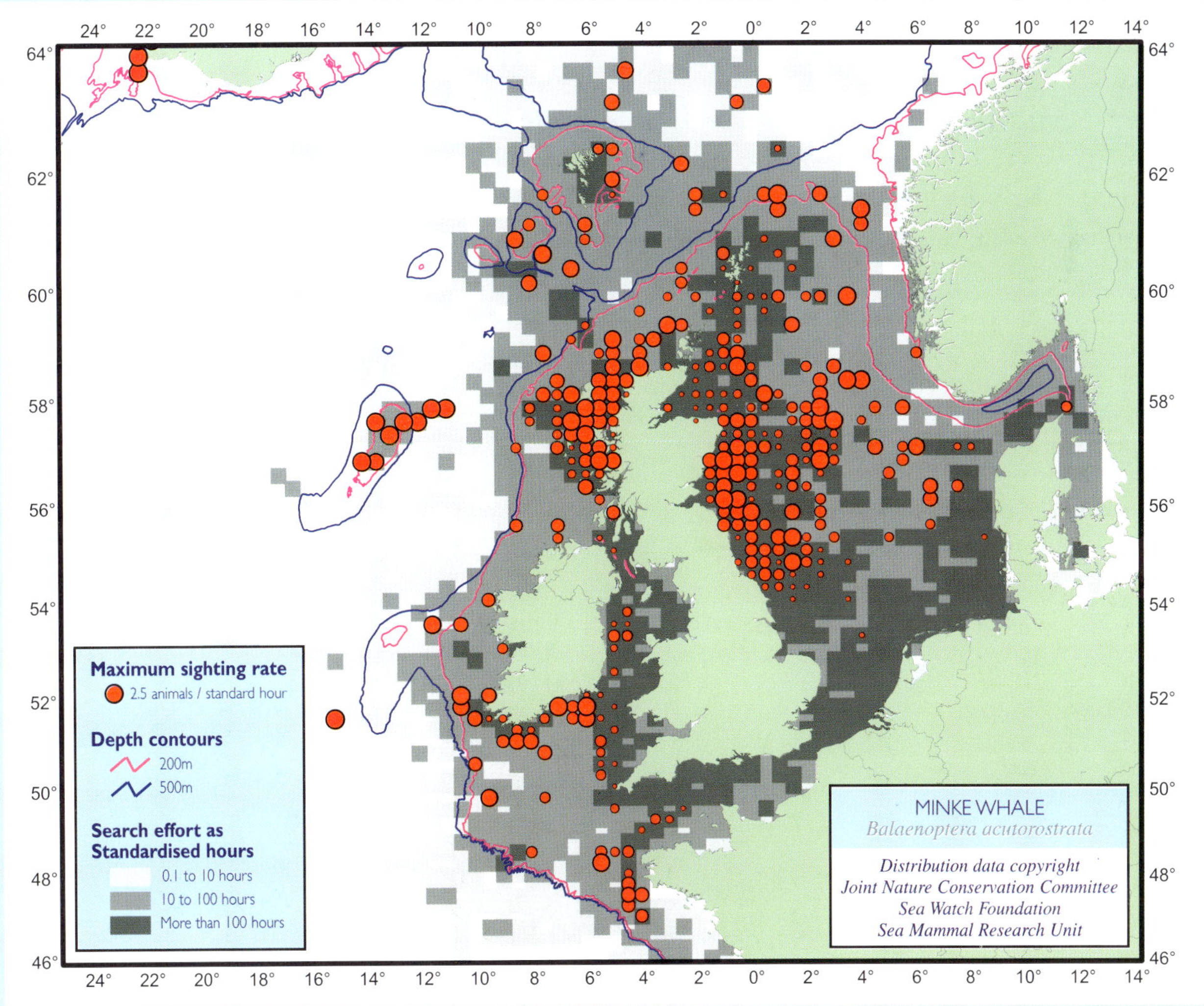

SEI WHALE

Balaenoptera borealis

The sei whale is a large baleen whale of the family Balaenopteridae that grows to a length of between 13.5 and 14.5 m. It has a similar appearance to the larger fin whale with which it can easily be confused, although it generally has a larger, more erect, and distinctly sickle-shaped fin; on surfacing, its 3 m high blow appears simultaneously with the fin before a shallow roll.

A relatively non-social species, sei whales are generally seen singly or in pairs, or otherwise in groups of up to five individuals. Larger aggregations up to 30 individuals have been recorded while feeding (Budylenko 1977). Some segregation by age and reproductive status may take place, and in the Southern Ocean, pregnant females migrate first, while younger animals rarely occur at the highest latitudes (Gambell 1985; Horwood 1987). Individuals sometimes breach clear of the water.

Sei whales feed primarily on surface plankton (mainly copepods, but also euphausiids), which they capture by skimming and swallowing, although small schooling fishes and squid form an important part of their diet in some areas, including the North Atlantic (Jonsgård and Darling 1977; Horwood 1987).

Global distribution

The sei whale has a worldwide distribution in mainly temperate and polar seas of both the northern and southern hemispheres. In the northern hemisphere, the species probably breeds mainly in warm temperate and subtropical waters during winter months, and then migrates northwards to summer in cold temperate and polar seas.

North Atlantic status

Summering populations are concentrated in deep waters (mainly 500-3,000 m depth) of the central North Atlantic north to Iceland (Horwood 1987; Sigurjónsson 1995). In the western north Atlantic, sei whales have been reported from two main locations, the Labrador Sea and Nova Scotia shelf, with winter records south to Florida (records in the Gulf of Mexico and Caribbean are questionable and may be misidentified Bryde's whale).

No current population estimates exist for the sei whale in the North Atlantic, but recent sightings surveys would indicate 13,500+ individuals, with evidence of depletion of stocks in some of the former whaling grounds (Sigurjónsson 1992; Cattanach *et al.* 1993).

NW European distribution

In the eastern North Atlantic, sei whales are thought to winter off north-west Africa, Spain and Portugal and in the Bay of Biscay, and then to migrate north to summering grounds off Shetland, the Faroes, Norway and Svalbard (Jonsgård and Darling 1977; Horwood 1987). Information on distribution patterns is constrained by confusion in the field with fin whale.

Sei whales apparently favour pelagic, temperate deep waters between 500 m and 3,000 m depth, with a more offshore distribution than fin whales or other balaenopterids (Horwood 1987; Sigurjónsson 1995). Most records from the British Isles come from waters deeper than 200 m between the Northern Isles and the Faroes (particularly in the vicinity of the Faroe-Shetland Channel and Faroese Bank Channel). Casual sightings of the species also occur occasionally in coastal waters off Shetland, the Outer Hebrides and between southern Ireland and south-west England (Evans 1992; Pollock *et al.* 1997, 2000). To the south, it is seen regularly in the Bay of Biscay in autumn and winter (Coles *et al.* 2001).

In UK waters, historical catches off the Outer Hebrides occurred mainly in June along the shelf edge near St Kilda; those taken in Shetland waters also came from the shelf edge, mainly in July and August. Recent sightings to the north and west of Scotland have mainly been in summer, between May and October but particularly in August (Pollock *et al.* 2000).

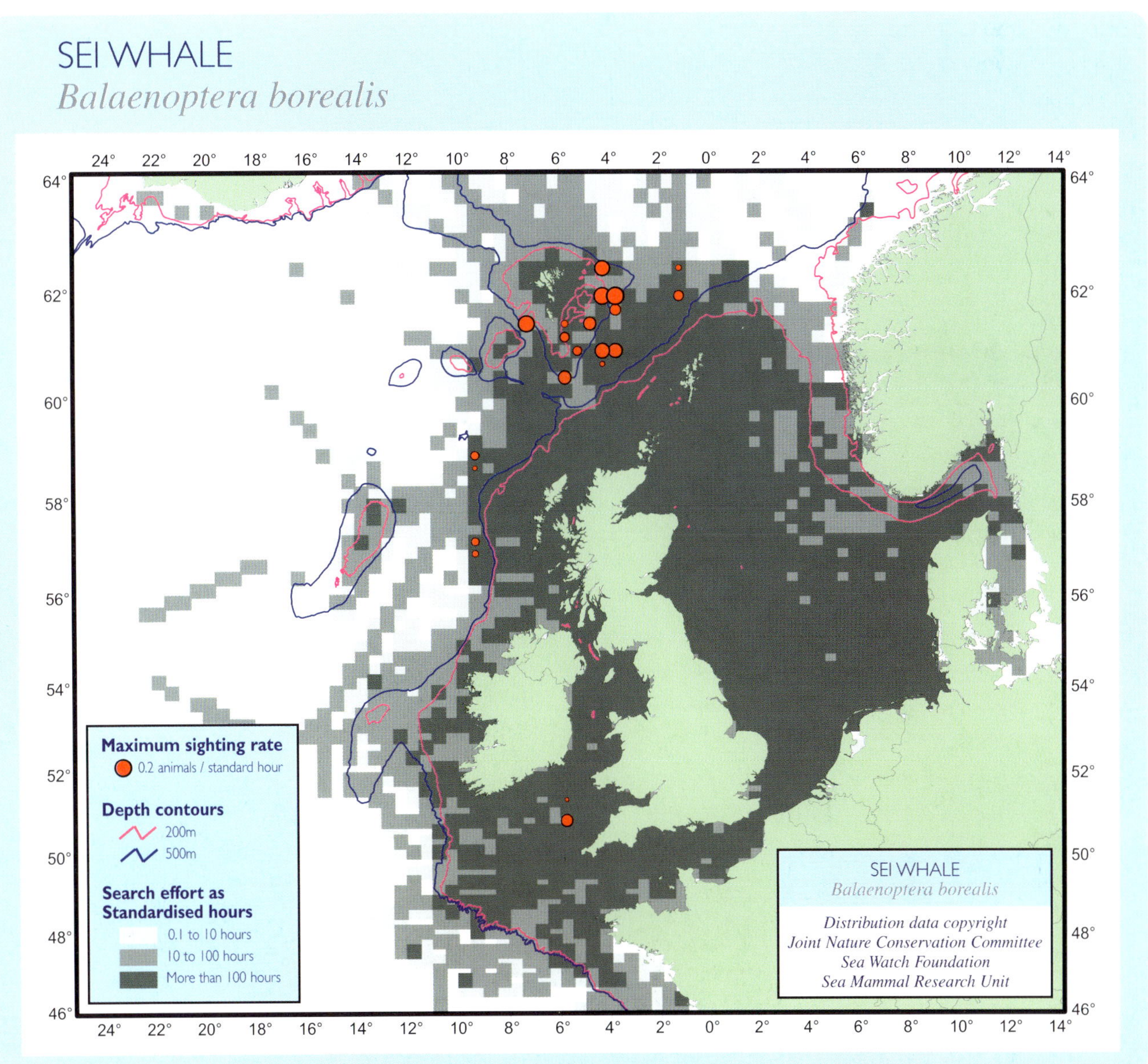

FIN WHALE

Balaenoptera physalus

The fin whale is the second largest of the baleen whales, reaching a length of between 17.5 and 20.5 m. In common with most other members of the family Balaenopteridae, it has a V-shaped head with a single central ridge and its dorsal fin is situated two-thirds along the back. Compared with the sei whale, with which it is often confused, the fin is small and points backwards. When it surfaces, the 4-6 m high blow is generally followed by a long shallow roll rather than the blow and dorsal fin showing simultaneously as in the sei whale. A diagnostic feature is the asymmetric patch of white extending on the right side over the lower lip, mouth cavity and front baleen plates.

A comparatively non-social species, most sightings of fin whales are of single animals or pairs. They do, however, form larger pods of 3-20 animals that may be part of a larger group, which in some parts of its range can number hundreds of individuals spread over a wide area, especially on the feeding grounds (Whitehead and Carlson 1988). Associations between individuals may change over a short time period, suggesting that these are fluid at least on the feeding grounds (Edds and MacFarlane 1987; Whitehead and Carlson 1988). The species is known to breach occasionally, and sexual behaviour involving excited chases has been observed at the surface (Stone *et al.* 1992).

Fin whales favour areas with high topographic variation such as underwater sills or ledges, and upwellings and frontal zones between mixed and thermally stratified waters with high zooplankton concentrations (Evans 1990; Relini *et al.* 1994b; Woodley and Gaskin 1996). The species feeds principally on planktonic Crustacea, mainly euphausiids but also copepods; they also take a variety of fish such as herring, capelin, sandeel, mackerel and blue whiting, and also cephalopods (Jonsgård 1966; Sigurjónsson 1995). A relationship between fin whale distribution and zooplankton abundance has been demonstrated in the western Mediterranean (Relini *et al.* 1998).

Global distribution

The fin whale has a worldwide distribution in mainly temperate and polar seas of both the northern and southern hemispheres. In the northern hemisphere, it probably breeds in warm temperate waters during winter months, and then migrates northwards to summer in cold temperate and polar seas. However, at least some individuals remain in high latitudes during winter. In the North Atlantic, for example, fin whales recorded by an extensive array of hydrophones are most vocally active north of 60° N between October and January (Charif and Clark 2000).

Fin whales live mainly in deep waters (400-2,000 m depth) beyond the edge of the continental shelf, but in some localities (e.g. the Bay of Fundy) they occur in shallow water less than 200 m deep.

North Atlantic status

Fin whales in the North Atlantic occur mainly from Baffin Bay west of Greenland south to Florida and the Gulf of Mexico in the west, and from Iceland and Norway south to the Iberian Peninsula in the east. A genetically distinct resident population inhabits the Mediterranean (particularly the Ligurian Sea - Bérubé *et al.* 1998).

Although there are no current population estimates for the North Atlantic as a whole, recent sightings surveys indicate a total population numbering 47,300 individuals, a figure smaller than historical estimates (Buckland *et al.* 1992; IWC 1992).

The fin whale was the most commonly taken large whale in the

Scottish and Irish fisheries between 1903 and 1928 (Brown 1976; Fairley 1991), and is the most frequently recorded whale from SOSUS hydrophone arrays (Charif and Clark 2000).

NW European distribution

In north-west Europe, the fin whale is distributed mainly along or beyond the 500 m depth contour - in the Norwegian Basin, Faroe-Shetland Channel, Rockall Trough, Porcupine Bight and the Bay of Biscay. Those localities where it has been recorded over the continental shelf tend to be close to the shelf edge, such as in northern Scotland and off southern Ireland. In the seas around the British Isles, fin whales occur mainly between June and December, although part of the population overwinters and breeds south of Ireland and in the Western Channel Approaches, where young calves and pregnant females have either been observed or recorded in strandings (Evans, 1992). During the early 20th century, catches in Scottish waters occurred mainly west and north of the continental shelf between April and October, with peak numbers north and west of Shetland during July and August (Thompson 1928; see also Evans 1990). Most sightings of fin whales in northern Britain occur between June and August, whereas in southern Britain and the Bay of Biscay, the species is observed year-round with peak frequency of sightings between June and November (Evans 1992; Pollock *et al.* 1997, 2000).

It has been suggested that fin whales undertake a general northwards movement off north-west Scotland from June to October (Stone 1997). However, vocalising animals have been recorded west of the European continental shelf between 42° and 62° N in all months of the year with no obvious seasonal latitudinal trend. Less vocalisation occurs from May through July (Charif and Clark 2000), perhaps because the animals are more vocal during the mating period in late autumn to winter (Boran *et al.* 2002).

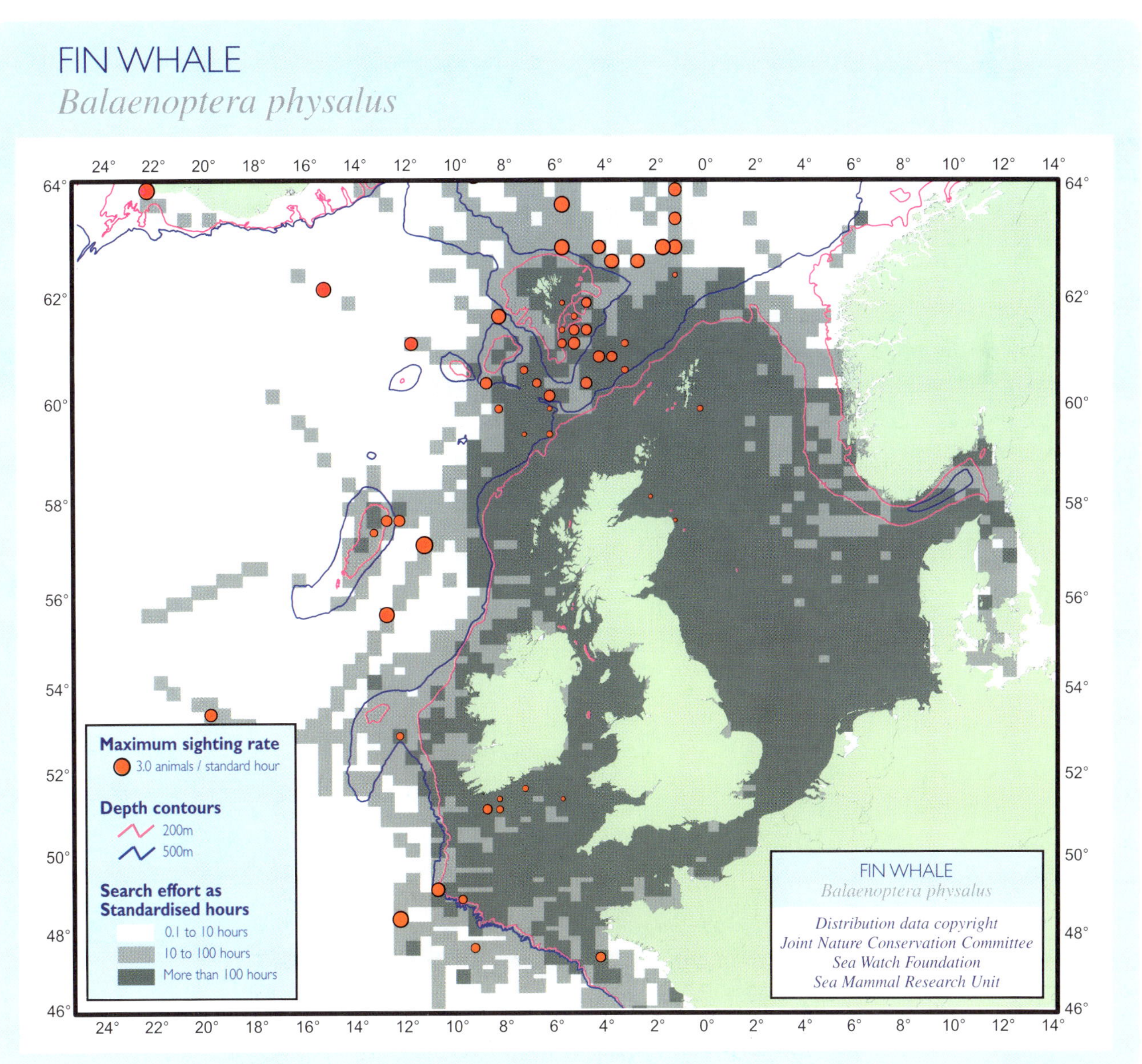

BLUE WHALE

Balaenoptera musculus

Adult blue whales range in length from 20-28 m. They have a flat U-shaped head, broader than either the fin or sei whale. The very small dorsal fin, varying from almost triangular to moderately sickle-shaped, is situated distinctly more than two-thirds along the back. Most of the body is pale bluish-grey, mottled with grey or greyish-white. When surfacing, the vertical, slender blow may reach a height of 9 m, followed by a long shallow roll. The fin is generally seen only just before the dive.

It is a relatively non-social species that is observed mainly singly or in pairs, but larger aggregations of up to five individuals occasionally occur in areas of high food concentration (Yochem and Leatherwood 1985). Mother-calf pairs have often been observed away from schools of males and non-reproductive females (Nemoto 1964). Blue whales raise their flukes before a long dive, and occasionally have been observed breaching. They are believed to be able to dive to 500 m, but probably exceed 200 m only rarely (Gambell 1979).

Blue whales feed largely on a few species of euphausiids, but also take copepods, and less frequently amphipods, cephalopods, and occasionally small fish; the latter may have been ingested accidentally (Kawamura 1980; Yochem and Leatherwood 1985). Feeding is usually by swallowing or gulping food, although side- and lunge-feeding, where the animal swims swiftly to the surface with mouth open, have been observed in the Gulf of St Lawrence and off California (Sears 1983).

Global distribution

Blue whales have a worldwide distribution in tropical, temperate and polar seas of both hemispheres. Like the sei and fin whales, blue whales probably breed mainly in warm temperate and subtropical waters during winter months, and then migrate northwards to summer in cold temperate and polar seas (Yochem and Leatherwood 1985).

North Atlantic status

In the North Atlantic, blue whales occur from the Caribbean in the west to the Canaries, Cape Verde Islands and West Africa in the east, and north to the Davis Strait/southern Greenland in the west across to Greenland and the Barents Sea/Svalbard in the east.

The North Atlantic population has been severely reduced by over-exploitation. Although no current estimates exist for the region as a whole, recent line-transect surveys would indicate a maximum of 442 whales (Gunnlaugsson and Sigurjónsson 1990; Sigurjónsson and Gunnlaugsson 1990) in the waters around Iceland at least. A long-term photo-identification study by Sears *et al.* (1990) has identified 203 individuals in the Gulf of St Lawrence, but it has not been possible to estimate overall abundance from this.

NW European distribution

Blue whales usually occur in deep waters (from 100-1,000 m), although in some regions they regularly occur close to land at depths of 200 m or less (Sears 1983; Sigurjónsson 1995). Off Iceland, they tend to occur closer to the coast than fin and sei whales (Sigurjónsson 1995).

Whaling records indicate that small numbers regularly passed west of Britain and Ireland in deep waters off the continental shelf (Thompson 1928), and more recently sightings (Evans 1992; Pollock *et al.* 1997, 2000) and acoustic monitoring (Charif and Clark 2000) have shown the species to occur in small numbers in deep waters of the Faroe-Shetland Channel and Rockall Trough and further south in the Bay of Biscay. Traditionally, blue whales were thought to winter in tropical and subtropical seas where they breed and then migrate to feed during summer months in cold temperate and polar waters.

However, recent acoustic monitoring using extensive arrays of hydrophones suggests that some remain in high northern latitudes throughout the winter months (Charif and Clark 2000).

RARER SPECIES (1)

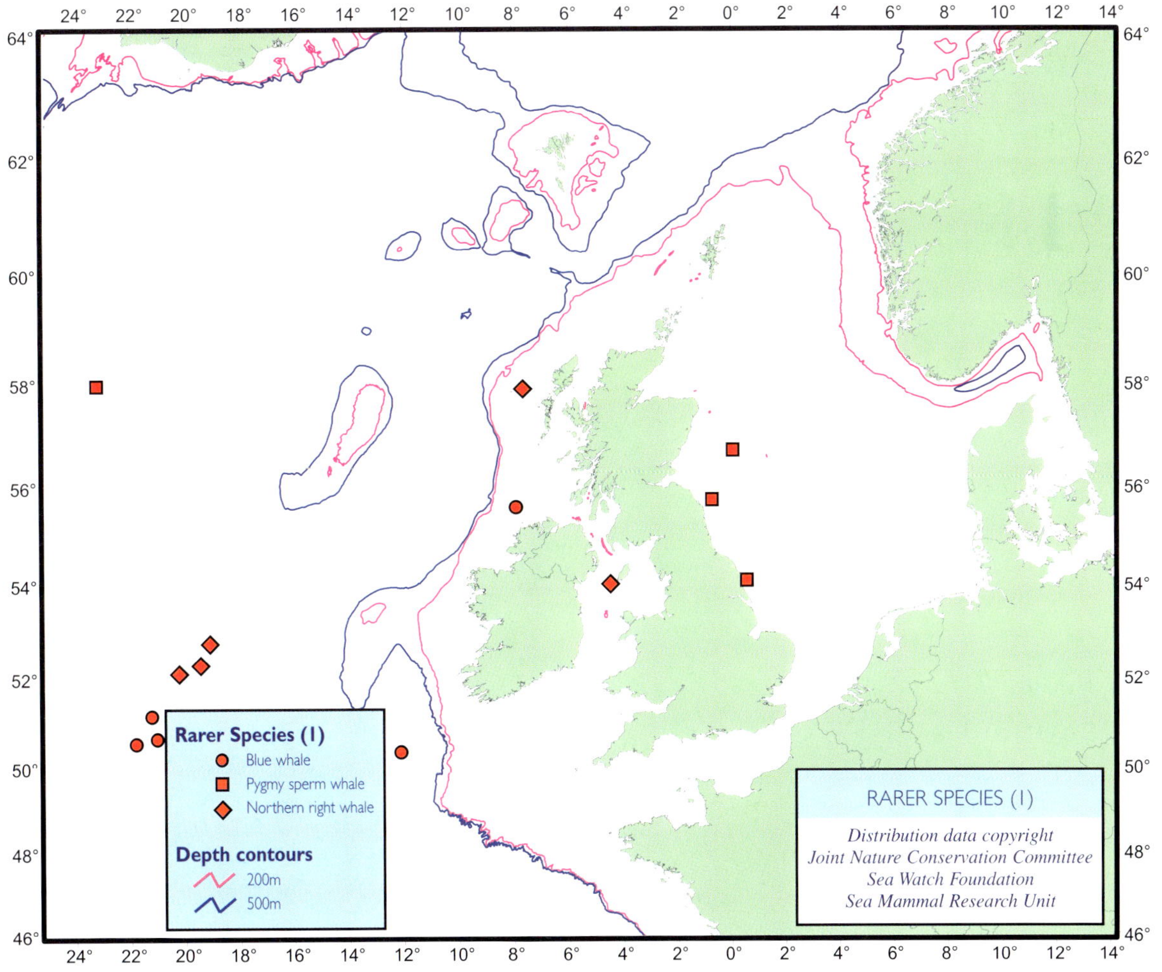

Rarer Species (1)
- Blue whale
- Pygmy sperm whale
- Northern right whale

Depth contours
- 200m
- 500m

RARER SPECIES (1)

Distribution data copyright
Joint Nature Conservation Committee
Sea Watch Foundation
Sea Mammal Research Unit

SPERM WHALE

Physeter macrocephalus

The sperm whale is the largest of the toothed whales, males growing to lengths of 18 m and females to 13 m. Its long, dark brown or grey back has no fin but there is a triangular, dorsal hump two-thirds along the body, followed by a spinal ridge and corrugations on the skin that give animals a shrivelled appearance. It has a very large, square head (up to one-third of the total length in the male) and an underslung lower jaw. The distinctive bushy blow is directed forwards and to the left, 1.5-5 m high. When diving deeply, the broad, triangular and deeply notched tail flukes are thrown up into the air.

The most social of the large whales, the sperm whale forms apparently matrilineal groups comprising adult females with their calves, and immature male and female offspring (Whitehead 1987; Rice 1989; Whitehead and Weilgart 2000). Females typically remain in the groups in which they were born; on reaching sexual maturity, males leave their natal group to join bachelor groups. Group size may number tens of animals, although they are usually spread over a large area, and only a proportion may be visible at the surface at any one time. About once a day, members of mixed groups come together at the surface into tight socialising/resting aggregations numbering 6-30 individuals (Whitehead 1987; Gordon 1998). Socially mature males associate briefly with matrilineal groups to breed. The spatial distribution of groups can be associated with the occurrence of high relief features, often over very large distances (Jaquet 1996).

Individual sperm whales, particularly juveniles, may breach clear of the surface, or lobtail (Waters and Whitehead 1990). Although most dives are apparently to depths of 400-600 m, they may dive as deep as 2,000 or even 3,000 m (Watkins *et al.* 1993).

Dives usually last 25-90 minutes, but exceptionally may last up to 138 minutes (Watkins *et al.* 1985; Papastavrou *et al.* 1989; Sarvas and Fleming 1999).

Sperm whale diet is varied but consists primarily of medium to large-sized mesopelagic squid (Kawakami 1980). Most prey have mantle lengths of 20 cm to 1 m, and in the North Atlantic comprise mainly members of the families Onychoteuthidae and Ommastrephidae, although giant *Architeuthis* and bottom-living octopus can also be taken. At high latitudes, such as off Iceland, deep-living fish including rays, sharks, red-fish, lumpsuckers, lantern fish, and gadoid fish are the predominant prey (Martin and Clarke 1986). Crustaceans are eaten very occasionally, and include giant mysids and benthic crabs (Kawakami 1980). Stomach contents analysis of animals stranded around the North Sea have revealed *Gonatus fabricii* to be by far the most important prey item, but a variety of other prey, including both deep-sea and coastal octopus, *Loligo forbesi*, *Histioteuthis bonnellii* and *Teuthowenia megalops*, also featured (Santos Vázquez *et al.* 1999).

Global distribution

The sperm whale has a worldwide distribution in tropical, temperate and subpolar seas of both the northern and southern hemispheres. It inhabits deep oceans (usually 500-2,000 m) from the equator to the edge of the polar pack ice.

Sperm whales are most commonly observed either in mid-ocean or over submarine canyons at the edges of the continental shelf, but can occur close to coasts of volcanic and oceanic islands in waters deeper than 200 m.

North Atlantic status

The species occurs in small numbers throughout deep waters of the North Atlantic, from the Davis Strait, Greenland and Norwegian Seas south to the Equator. It also occurs throughout the Mediterranean Sea. Only males move into sub-Arctic waters to feed, whereas females and immatures tend to occur in more discrete areas at lower latitudes. There are no reliable population estimates for sperm whales in the North Atlantic, but hunting in the past is believed to have reduced population size (Rice 1989). Nevertheless, the species is common in deep waters. Major breeding areas include the Caribbean Sea and around the Azores.

NW European distribution

Male sperm whales in north-west Europe occur mainly in waters deeper than 200 m around Iceland, west of Norway, beyond the shelf break north and west of Scotland and Ireland and in the Bay of Biscay. They have also been observed in near-shore waters mainly off Iceland, western Norway, and the Northern Isles of Scotland. Both males and females have been observed off the west coast of Portugal, the north coast of Spain, and around the Azores (Evans 1997; Gordon 1998).

Sightings in British and Irish waters have been recorded mainly between July and December; there is increasing evidence that small groups of males remain at high latitudes into winter, when mass-strandings have taken place (Berrow *et al.* 1993; Evans 1997). During the Scottish and Irish whale fisheries of the early 20th century, most sperm whales were caught between June and August (Thompson 1928; Fairley 1981).

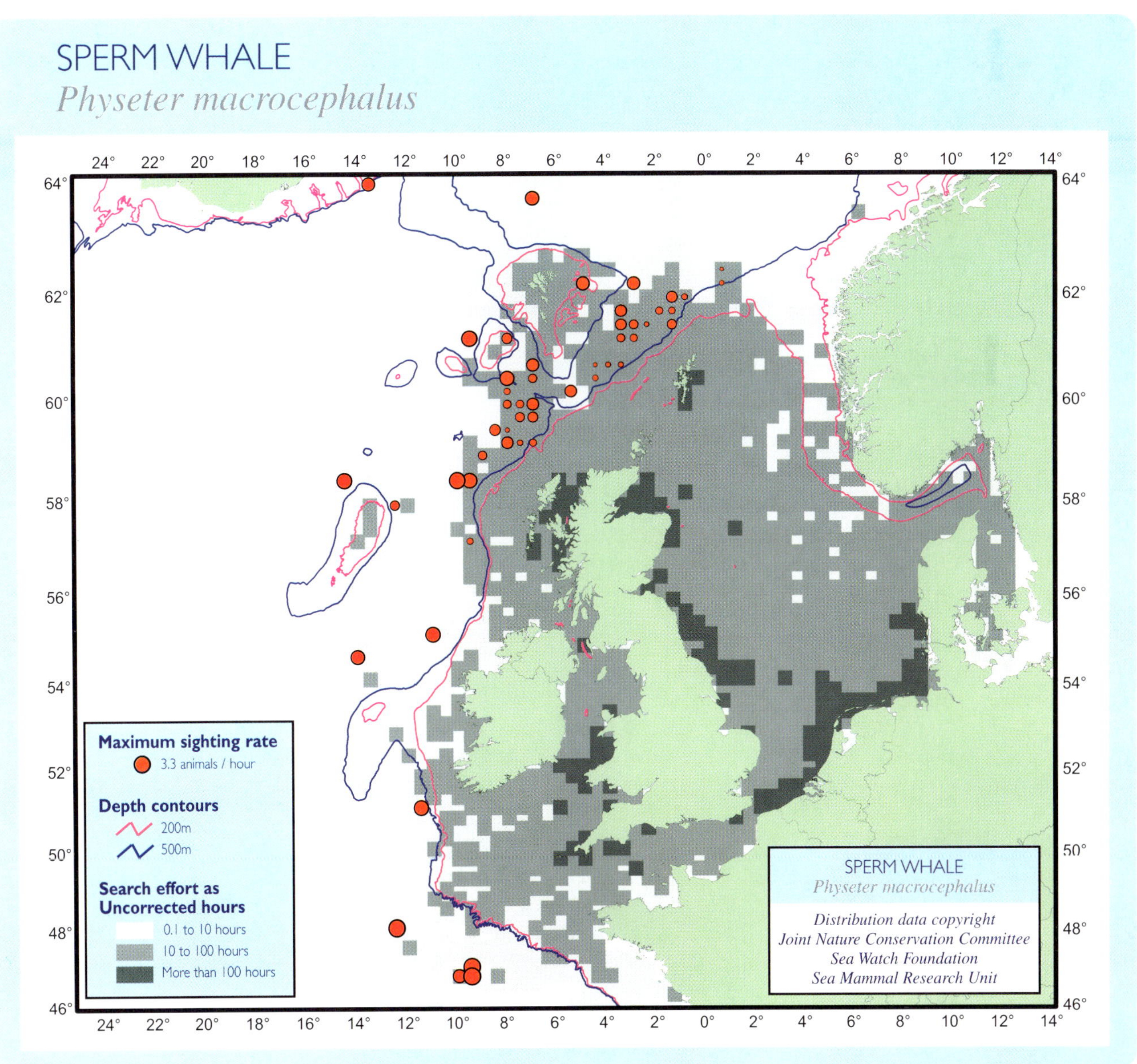

PYGMY SPERM WHALE

Kogia breviceps

The pygmy sperm whale looks superficially like a dolphin. It is around 3 m in length, and has long tapering flippers and a low, strongly sickle-shaped dorsal fin situated just behind the centre of the back. In contrast to various dolphin species, however, it rises slowly to breathe and is a relatively slow swimmer. A close view reveals a squarish or conical, shark-like head with underslung lower jaw. On each side of the head behind the eye is a crescent-shaped pale mark contrasting with a dark line behind it, giving the appearance of a shark's gill.

A poorly known species, observations of pygmy sperm whales have been mainly of single individuals or small groups numbering up to six animals. It is apparently easy to approach, although it rarely comes near vessels (Caldwell and Caldwell 1989). It occasionally breaches, but more usually is observed lying motionless at the surface, at least in calm weather, in the manner of a sperm whale.

The diet is predominantly mesopelagic squid, although there is some evidence that fish and crustaceans are also taken (Martins *et al.* 1985; McAlpine *et al.* 1997; Plön *et al.* 1999). Feeding studies indicate that females with calves take more inshore species, whereas adult males consume cephalopods that inhabit the continental shelf edge and slope (Plön *et al.* 1999).

Global distribution and North Atlantic status

The pygmy sperm whale has a worldwide distribution in tropical, subtropical and temperate seas of both the northern and southern hemispheres.

It occurs mainly in deep oceanic waters beyond the edge of the continental shelf.

The distribution and abundance of the pygmy sperm whale in the North Atlantic is poorly known. Most information comes from strandings, the majority of which have occurred on the coasts of North America. Records range from Nova Scotia in the west and Britain and Ireland in the east, south to Dominica in the west and the Canary Islands in the east (Caldwell and Caldwell 1989).

NW European distribution

There are very few records of this species from European waters. There have been strandings in Spain (Nores and Pérez 1982; Penas-Pantiño and Piñeiro Seage 1989), Portugal (Reiner 1981; Sequeira *et al.* 1992), the Atlantic coast of France (Duguy 1966), the Netherlands (Van Oort 1926), Wales (Sheldrick 1989) and western Ireland (Fraser 1974; Berrow and Rogan 1997). The few sightings are mainly from the Bay of Biscay, the South-west Approaches to the Channel and western Ireland, although the species has occasionally been recorded in the North Sea (Evans 1992; A.D. Williams, pers. comm.).

CUVIER'S BEAKED WHALE

Ziphius cavirostris

Cuvier's beaked whale has a long stout body with a small head, short indistinct beak, and concave or slightly S-shaped mouth line (likened to a goosebeak) giving rise to a slightly protruding lower jaw. It grows to a length of c. 7 m, the female being slightly larger than the male. The body is usually brownish grey; the head and front portion of the back lightens with age, and can even be white in older animals.

The limited number of sightings at sea of Cuvier's beaked whale have mostly been either of single individuals or small groups of two to seven individuals (Leatherwood and Reeves 1983; Heyning 1989). No information is available on social organisation, although in the Bay of Biscay, adult groups seem to contain only one male, and groups containing juveniles include only adult females (Coles *et al.* 2001). The dorsal surface and flanks of males often show extensive scarring, thought to be due to intra-specific aggression (Montero and Martín 1992).

Although they were once thought to be wary of boats (Leatherwood and Reeves 1983), recent observations have belied this, and the dearth of sightings at sea is probably a result of the species' deep-diving habit and inconspicuous surfacing behaviour; dives can last for at least 30 minutes. Breaching has also been observed. The species has been recorded in association with other Ziphiidae, including northern bottlenose whales and Gervais' beaked whales (Vonk and Martin 1989).

A preference for deep water is reflected in the species' diet of mainly deep-water cephalopods. Stranding data show that a wide variety of squid are taken, along with various species of fish including gadoids and blue whiting, and crustaceans (Clarke 1986; Heyning 1989; Santos Vázquez *et al.* 2001).

Global distribution

Cuvier's beaked whale has the widest geographical range of all beaked whales, occurring from the tropics to cold-temperate regions of all oceans. Most information on its distribution has come from strandings, which suggests that the species typically inhabits tropical waters, moving into temperate seas during the warmer summer months. It is a pelagic species, mainly found in waters between 500 and 3,000 m depth.

North Atlantic status

No population estimates exist for Cuvier's beaked whale. The distribution of both sightings and strandings suggests that it is most commonly associated with the warm waters of the Gulf Stream. Strandings are known from the east coast of the United States from Nova Scotia south to Florida and the Caribbean (Heyning 1989; Houston 1991). The species has been recorded frequently off the Iberian Peninsula, in the Bay of Biscay and in the Mediterranean Sea (Castells and Mayo 1992; Cagnolaro *et al.* 1983; Coles *et al.* 2001), and also off the Azores (Reiner *et al.* 1993) and the Canary Islands (Montero and Martín 1992).

NW European distribution

There have been only six confirmed sightings of Cuvier's beaked whale in British and Irish waters, although strandings are not uncommon. Sightings occurred east of the Orkney Islands in the northern North Sea in August 1980, off southern Ireland in August 1984, off south-west Ireland in July 1987, off north-west Ireland in August 1987, off the Outer Hebrides of Scotland during September 1988, and off southern Ireland in June 1998 (Evans 1992; Rosen *et al.* 2000). The occurrence of all sightings between June and September would suggest a summer movement of animals into UK waters. Further south in the Bay of Biscay, the species may be resident year-round (Coles *et al.* 2001). The species has stranded mainly along the western seaboard of Britain and Ireland, and only rarely in the North Sea.

NORTHERN BOTTLENOSE WHALE

Hyperoodon ampullatus

Growing to a length of up to 9.5 m, the northern bottlenose whale is recognisable by its short, dolphin-like beak, large, bulbous head (flatter in old males), and erect, hooked fin situated two-thirds along the back. On surfacing, it has a bushy blow, 2 m high, and slightly forward-pointing. The chocolate brown or dark grey upper parts become paler with age.

Northern bottlenose whales are most frequently observed singly or in groups of 2-7, but occasionally they may form groups of up to 35 animals (Evans 1980; Bloch *et al.* 1996). Groups can occur in close proximity to one another (Benjaminsen and Christensen 1979), and early in the twentieth century, Nansen (1925) reported aggregations of several hundred, thought to be on migration. Social organisation in the species is poorly known but the composition of groups appears to be variable; groups comprising mature females with juveniles, immature males only, mature males only, and females, immatures and mature males together have all been recorded (Faucher and Whitehead 1991; Bloch *et al.* 1996). Groups of mixed sex are more commonly observed between June and August (Whitehead *et al.* 1997).

Northern bottlenose whales are deep divers and can reach depths of 1,450 m in dives lasting up to 70 minutes (Hooker and Baird 1999). After such long dives they may linger at the surface for 10 minutes or more. Care-giving behaviour has been reported frequently (Benjaminsen and Christensen 1979; Mead 1989b). Bottlenose whales frequently approach vessels and have been observed to lobtail and, occasionally, breach.

The diet is dominated by squid of a variety of species (particularly *Gonatus fabricii* and species of the genera *Histioteuthis* and *Octopoteuthis*), but it also includes fish (for example, herring, Greenland halibut, lumpsucker, redfish, ling, skate and spiny dogfish), large decapod crustaceans (*Pandalus* spp.), sea stars, and sea cucumbers (Benjaminsen and Christensen 1979; Clarke and Kristensen 1980; Bloch *et al.* 1996; Lick and Piatkowski 1998; Santos Vázquez *et al.* 2001).

Global distribution and North Atlantic status

Northern bottlenose whales occur only in temperate, subpolar and polar seas in the North Atlantic. The distribution ranges from the Barents Sea, southern Greenland and Svalbard southwards as far as latitude 15° N. They are more numerous at higher latitudes, seeming to favour cold water between 0° C and 2.5° C, and they occasionally enter pack ice off Svalbard and Labrador (Reeves *et al.* 1993). Animals often occur along the boundaries between cold polar and warmer Atlantic currents.

North Atlantic sightings surveys (NASS) in 1987 and 1989 suggested a northern bottlenose whale population numbering about 40,000 animals (Víkingsson 1993; NAMMCO 1993). Hunting of the species between the 1880s and 1970s probably contributed to a large population decline (Klinowska 1991). However, the species does appear to remain locally abundant at some locations, for example the Gully, Nova Scotia and south and east of Iceland (Gunnlaugsson and Sigurjønsson 1990; Whitehead *et al.* 1997).

NW European distribution

In north-west Europe, bottlenose whales are most frequently recorded in deep water. Most sightings have been made north and west of Scotland, mostly along the continental shelf edge over the 1,000 m isobath, and also around the Faroe Islands. The species appears less common west of Ireland and is rarely seen in shelf waters of the Channel, Irish Sea or North Sea. However, it does occur regularly further south, in the deep canyons (*c.* 3,000 m) of the southern Bay of Biscay (Carlisle *et al.* 2001).

It has been suggested that bottlenose whales undergo latitudinal migrations, moving northwards in spring and southwards in autumn (Thompson 1928; Jonsgård and Øynes 1952).

In summer, the species appears to move towards NW European shelf waters, where most records occur between April and September; peak sightings in the Bay of Biscay occur between June and August (Coles *et al.* 2001: Carlisle *et al.* 2001), off northern Scotland in August (Evans 1992; Sea Watch, unpubl. data), and in the Faroes in August and September, when catches are also highest (Bloch *et al.* 1996). However, a small number of animals are caught throughout the year off the Faroes, so at least part of the population winters at high latitudes.

NORTHERN BOTTLENOSE WHALE

Hyperoodon ampullatus

Maximum sighting rate

14.2 animals / hour

Depth contours

200m

500m

Search effort as Uncorrected hours

0.1 to 10 hours

10 to 100 hours

More than 100 hours

NORTHERN BOTTLENOSE WHALE

Hyperoodon ampullatus

Distribution data copyright
Joint Nature Conservation Committee
Sea Watch Foundation
Sea Mammal Research Unit

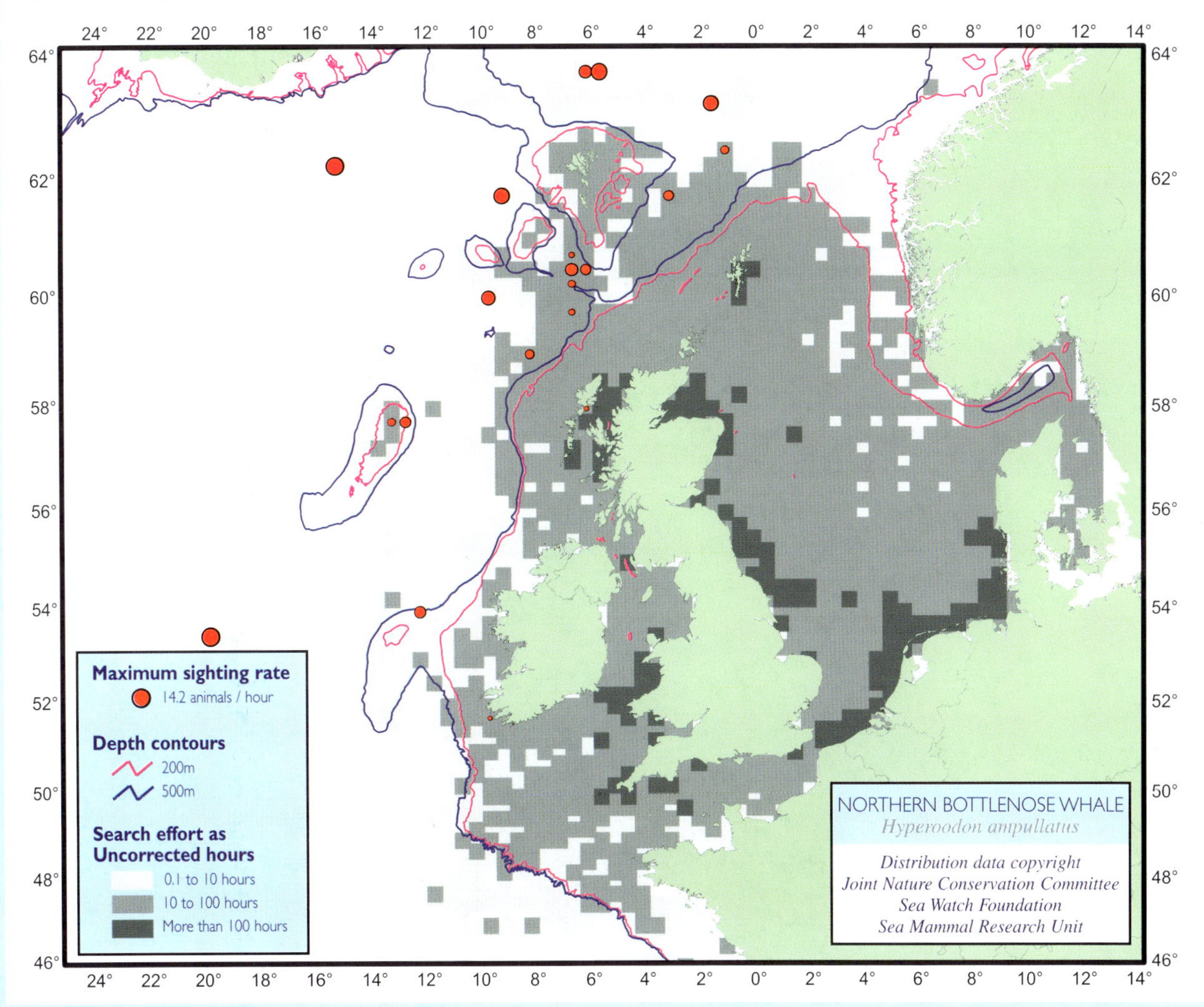

BEAKED WHALES

Mesoplodon spp.

Beaked whales of the genus Mesoplodon *are relatively infrequently encountered and are therefore rather poorly known. They are difficult to identify because of their inconspicuous surfacing behaviour and the similarity between species (distinguished reliably primarily by the size and position of the single pair of flattened triangular teeth, protruded only in adult males); consequently, many animals observed at sea have not been identified to species level. They grow up to 6.5 m long and have a laterally compressed, spindle-shaped body with a small head, often a prominent bulge in front of the blowhole, and a well-developed beak.*

Mesoplodonts have a global distribution, but appear to be absent from Arctic and Antarctic waters. They have been observed breaching and tail-slapping, as well as porpoising, but generally their behaviour is unremarkable when surfacing (Ritter and Brederlau 1999). The diet consists of deep-water species of squid and occasionally fish (Mead 1989a).

Four species have been recorded in north-west Europe: Sowerby's beaked whale *M. bidens*, True's beaked whale *M. mirus*, Gervais' beaked whale *M. europaeus*, and Blainville's beaked whale *M. densirostris*. Sowerby's beaked whale has a long, slender beak and moderately arched lower jaw, with (in adult males, as with all mesoplodonts) one pair of teeth slightly behind the mid-point of the gape. True's beaked whale has a short but clearly defined beak sloping into a slightly bulbous forehead, with a single pair of teeth at the extreme tip of the lower jaw.

Gervais' beaked whale too has a short beak, and relatively straight mouthline, with one pair of teeth in the lower jaw about one-third along the gape from the tip of the snout. Blainville's beaked whale has a high, arching prominence near the corner of the mouth giving the species a prominent contour, at the apex of which a massive pair of teeth is exposed, tilting slightly forward and often encrusted with barnacles. Unlike the other three species which are more uniformly grey in colour, it commonly has large pale blotches over the back. The head is often flattened directly in front of the blowhole, from which protrudes a moderately long, slender beak (in Gervais' beaked whale, the rostrum is ventrally flattened).

Sowerby's beaked whale

Sowerby's beaked whale has the most northerly distribution of all species of *Mesoplodon* in the Atlantic, and is the most frequently seen and stranded species in the north-east Atlantic. Social structure is poorly known; usually, it is recorded alone or in pairs, but groups of up to ten and of mixed-sex have been seen in the north-west Atlantic (Hooker and Baird 1999).

Sowerby's beaked whale. J Benney.

BEAKED WHALES
Mesoplodon spp.

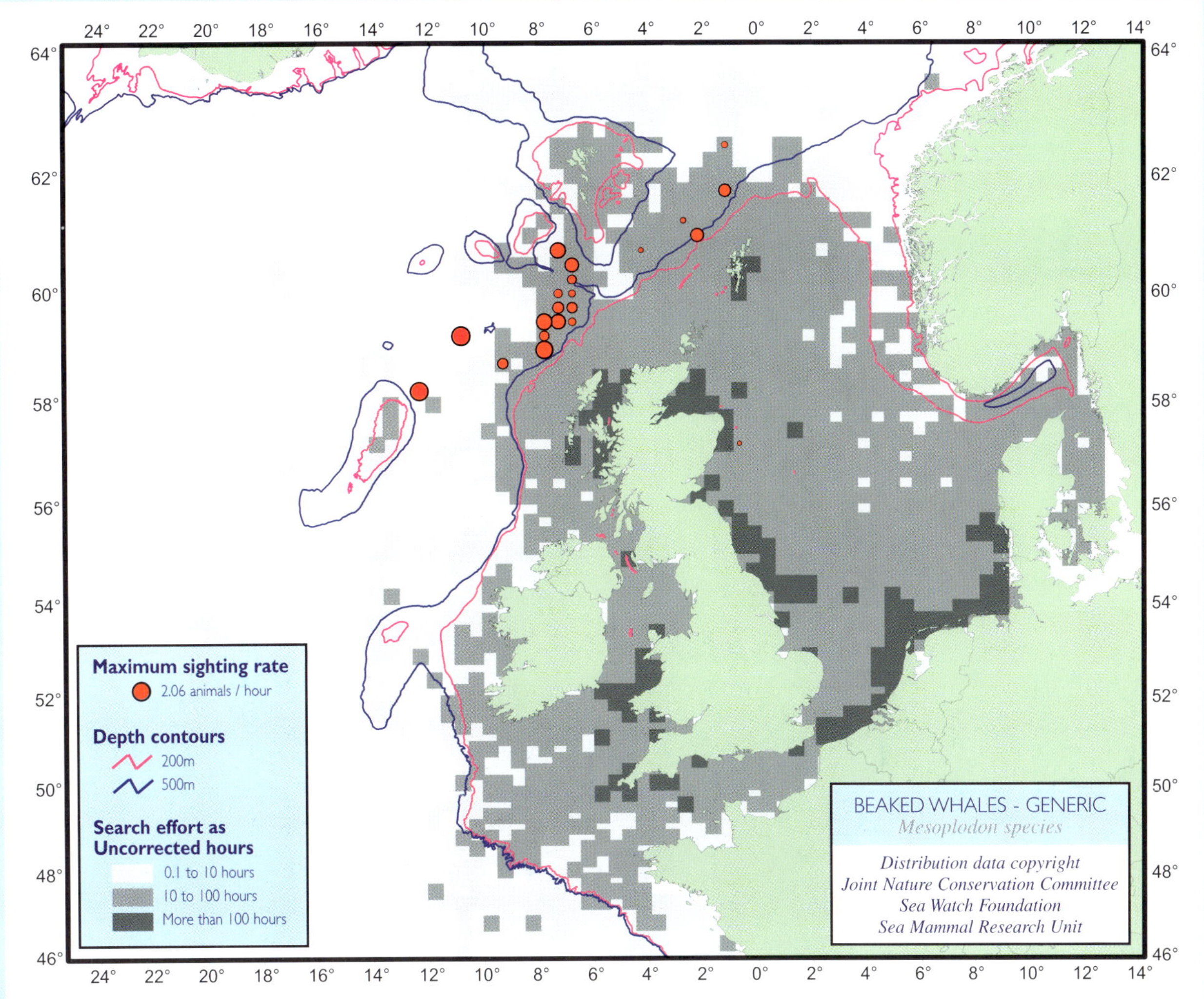

Maximum sighting rate
2.06 animals / hour

Depth contours
200m
500m

Search effort as Uncorrected hours
0.1 to 10 hours
10 to 100 hours
More than 100 hours

BEAKED WHALES - GENERIC
Mesoplodon species

Distribution data copyright
Joint Nature Conservation Committee
Sea Watch Foundation
Sea Mammal Research Unit

RARER SPECIES (2)

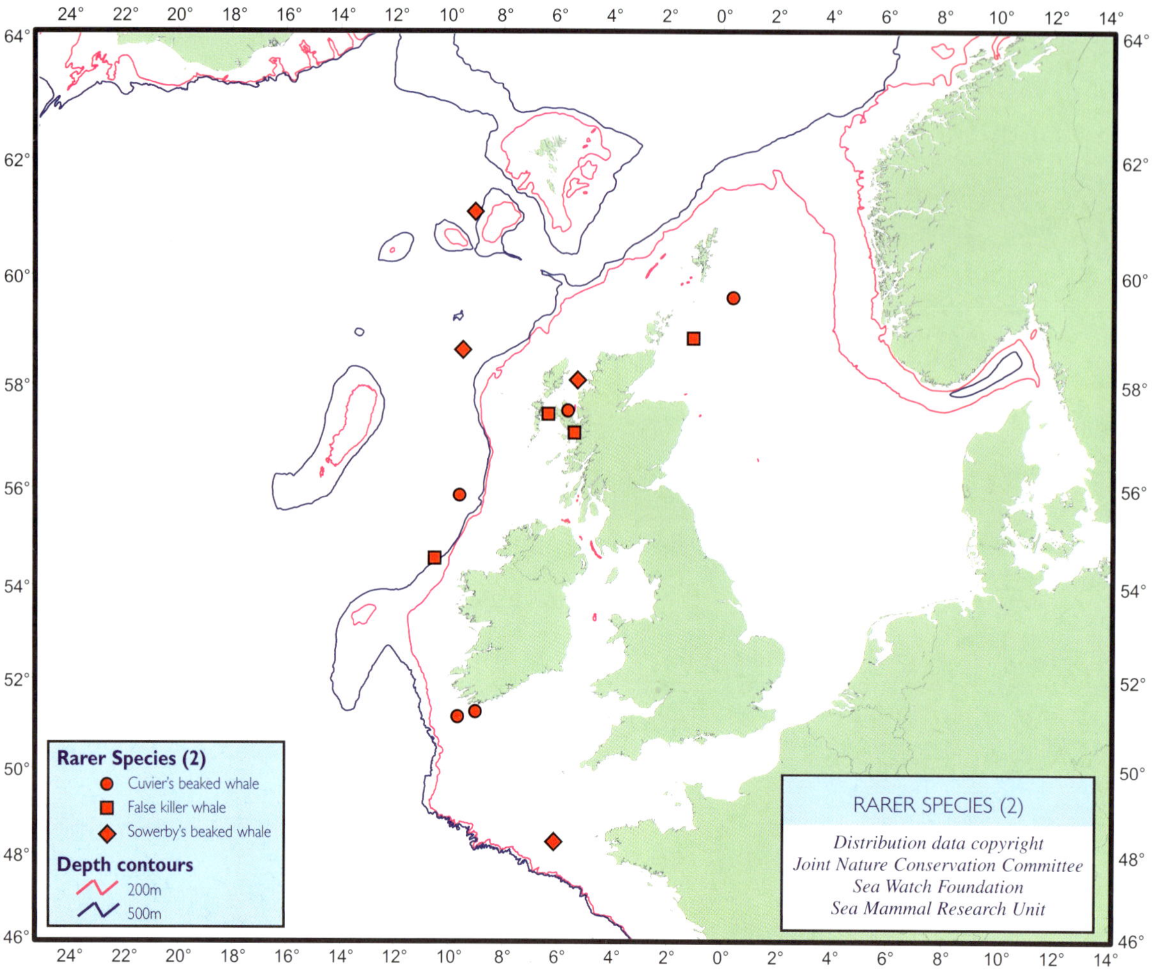

Rarer Species (2)
Cuvier's beaked whale
False killer whale
Sowerby's beaked whale
Depth contours
200m
500m

RARER SPECIES (2)
Distribution data copyright
Joint Nature Conservation Committee
Sea Watch Foundation
Sea Mammal Research Unit

Sowerby's beaked whale. J Benney.

It appears to prefer deep (greater than 1,000 m) water trenches and has been recorded south of Iceland, in the Norwegian Sea, west of Norway, around the Faroe Islands, north and west of Britain and Ireland and in the North Sea, South-west Approaches to the Channel and the Bay of Biscay (Lien and Barry 1990; Evans 1992; Carlstrom *et al.* 1997; Bloch 1998; Coles *et al.* 2001). The species has also been observed off Madeira (Young *et al.* 1999) and regularly in the Azores (Reiner 1986; E. Simas, pers. comm.).

True's beaked whale

This species inhabits warm-temperate seas, mainly in the North Atlantic and in the southern hemisphere off South Africa and Australia (McCann and Talbot 1963; Ross 1969; Mead 1989a). There are very few confirmed sightings of True's beaked whale at sea in the north-east Atlantic and the species is clearly less common here than Sowerby's beaked whale. Strandings on European coasts have occurred mostly along the west coast of Ireland. As these show no seasonal peaks, the species may occur off Britain and Ireland throughout the year. Elsewhere in the north-east Atlantic, it has been reported in the Bay of Biscay (Walker *et al.* 2001) and off the Azores (Steiner *et al.* 1998), but is more frequently seen in the north-west Atlantic, where records range from Nova Scotia to the Bahamas (Mead 1989a).

Gervais' beaked whale

Gervais' beaked whale probably favours warm temperate and tropical Atlantic waters. The majority of records come from the western Atlantic, where it has been observed between New York and the Caribbean. There have been very few strandings (only one in Britain and Ireland, in Co. Sligo in January 1989) and confirmed sightings in the north-east Atlantic, a sighting of three in January 1998 being the most recent (Carrillo and Martín 1999).

Blainville's beaked whale

Blainville's beaked whale is probably the most widely distributed species of *Mesoplodon*, and occurs in the tropical and warm temperate waters of all oceans. The species is much more frequently recorded in the north-west Atlantic, particularly the Gulf of Mexico, the Caribbean and the east coast of the United States between Florida and Massachusetts (Mead 1989a). There are few records of Blainville's beaked whale from the north-east Atlantic. Extralimital strandings have occurred in Iceland (A. Petersen, pers. comm.) and Britain (West Wales, July 1993 - Herman *et al.* 1994); other strandings have been mainly in the Iberian peninsula (e.g. Sequeira *et al.* 1996; Valverde and Galán 1996) as well as in Madeira (Harmer 1924) and the Canary Islands (Carrillo and Lopez-Jurado 1998). In the Canary Islands, most live sightings at sea have been in water depths of between 100 and 500 m, where mean and maximum recorded group sizes were 3.4 and 9 animals respectively (Ritter and Brederlau 1999).

Within the study area, all records of unidentified *Mesoplodon* species were from the north and west of Scotland, mostly of single or pairs of animals, but groups comprising as many as ten animals were recorded. A cluster of sightings to the north-west of the Outer Hebrides at around 60° N and 8° W is over varied seabed topography, where sea depths vary between 500 m and 1,400 m; complex currents may produce a rich food supply here. Virtually all such sightings occurred in water approaching or deeper than 1,000 m.

Blainville's beaked whale. Unknown.

The shallow water in the North Sea off north-east Scotland in which a single, unidentified beaked whale was recorded, is probably an atypical habitat for beaked whales in north-east Atlantic waters.

Sightings of unidentified mesoplodonts were made in most months. This suggests that they remain in northern and western waters off Britain and Ireland throughout the year, albeit with a peak in numbers in August. Records of juvenile animals also suggest that these waters are used for both breeding and feeding. Although specific identification of these animals was not possible, it seems likely from field descriptions and stranding data that most of them have been Sowerby's beaked whales.

BELUGA OR WHITE WHALE

Delphinapterus leucas

The adult beluga is all white; young animals are slate grey to reddish-brown, changing to blue-grey at two years old. The beluga has a stout body, a small head, well-defined neck, usually prominent rounded melon, and a short but distinct beak. The flippers are short, broad and paddle-shaped, but most distinctively, the species lacks a dorsal fin, the narrow-ridged back darkening slightly along the top. Sometimes old pale Risso's dolphins or bleached dead pilot whales are mistaken for white whales, but confusion should not exist, since those species have a noticeable dorsal fin.

The beluga is a gregarious species, usually living in groups of 2-20 that often are segregated by age and reproductive status, but which may aggregate seasonally to number hundreds or even thousands of animals. The basic social unit is the mother-calf pair, with sometimes the previous calf also in attendance. It is a deep diver, regularly reaching depths of hundreds of metres and occasionally more than 1,000 m (Martin 1996; Martin *et al.* 1998). The diet includes a wide variety of fish, as well as squid, octopus, decapod crustaceans, molluscs, and annelid worms (Seaman *et al.* 1982; Heide-Jørgensen and Teilman 1994).

Global distribution and North Atlantic status

The beluga is restricted to Arctic and sub-Arctic seas of the northern hemisphere. A vagrant to UK waters, its distribution is centred upon areas such as Baffin Bay, and the Greenland and Barents Seas. It lives in cold waters near or within sea ice and may enter estuaries and river mouths for short periods in summer, possibly associated with its annual synchronised moult (Martin 1996).

NW European distribution

During the 20th century there were 15 records of beluga in UK waters, all adults and all live sightings except for one (Evans 1992; Sea Watch Foundation, unpubl. data). These occurred off north-west Scotland, around the Northern Isles and in the North Sea. Some sightings were probably of the same individuals, perhaps originally part of a larger group, as in several instances there were records from neighbouring North Sea countries around the same times (Evans 1992).

NARWHAL

Monodon monoceros

The narwhal has a stout body with a small, rounded head, bulbous forehead and very slight beak, and no dorsal fin. In the upper jaw there is a single pair of teeth, the left tooth of which in the male erupts through the upper lip and becomes greatly extended to form a spiralled tusk up to 3 m in length. Narwhals are gregarious and are rarely seen alone; they associate in groups of 3-4, sometimes up to 20 individuals. Aggregations of several hundreds or even thousands animals may occur seasonally (Reeves and Mitchell 1987; Kingsley et al. *1994). Within these, tightly-knit groups occur usually of animals of similar age and sex. The frequent scarring of adult males suggests that aggressive behaviour between them is common.*

Narwhals are slow swimmers, averaging only 5 km/hr during migration (Dietz and Heide-Jørgensen 1995). They are also deep divers, regularly reaching depths of 500 m when feeding, but capable of exceeding 1,000 m (Heide-Jørgensen and Dietz 1995). The diet of narwhals includes a variety of prey including fish, squid and shrimps. Species taken include Arctic cod, polar cod, Greenland halibut, red-fish, decapod and euphausiid crustaceans, squid and octopus (Finley and Gibb 1982; Heide-Jørgensen *et al.* 1994).

Global distribution and North Atlantic status

The distribution of the narwhal is restricted almost entirely to Arctic seas of the northern hemisphere, such as Baffin Bay and the Greenland and Barents Seas. It lives mainly in deep offshore waters near or within sea ice, in summer coming into fjords and bays (occasionally at depths of less than 10 m; Martin, in press).

The species occurs in three main areas: the Canadian high arctic east into Baffin Bay and the northern Davis Strait; northern Hudson Bay, Foxe Basin and Hudson Strait; and from East Greenland across the Greenland Sea to Svalbard, Franz Josef Land and east to the Kara Sea (Reeves *et al.* 1994). No complete abundance estimates are available for any of these areas. In West Greenland, where narwhals undergo a seasonal movement south, up to 4,000 have been counted in August in the north-west (Inglefield Bredning and Melville Bay) and 3,000 in Disko Bay in March. From late November to May, 35,000 have been estimated offshore in the northern Davis Strait and Baffin Bay. In the northern Hudson Bay and Foxe Basin, an apparently isolated group of 1,300 persists. Low numbers exist offshore in the Eurasian sector of the Arctic Ocean, and the species probably occcurs in low numbers in the Greenland Sea and along the East Greenland coast (Reeves *et al.* 1994).

NW European distribution

The narwhal is more strictly a high Arctic species than is the beluga and is a very rare vagrant to UK waters. During the last hundred years, there have been only three confirmed records of narwhal in the UK, all strandings in 1949 (Fraser 1974).

COMMON BOTTLENOSE DOLPHIN

Tursiops truncatus

Growing to a length of 4 m in the north-east Atlantic (though larger elsewhere), the bottlenose dolphin is a large dolphin with a robust head and distinct short beak, which is often white-tipped on the lower jaw. It has a brown or dark grey back, lighter grey lower flanks grading to white on the belly, with no obvious flank markings. It has a centrally-placed, fairly tall and slender dorsal fin, which is usually sickle-shaped; individually unique nicks and markings on the fin and back are often used to identify individuals.

It is a social dolphin, commonly forming groups of 2-25, but occasionally numbering several tens or low hundreds of animals. Larger schools tend to occur in deeper waters, where in some parts of the world distinct offshore forms have been recognised. School size may vary through the day, according to age or sex class (females with calves usually occur in larger schools than males), and may also depend on activity. However, many associations are relatively stable (Wells *et al.* 1987; Smolker *et al.* 1992; Connor *et al.* 1992). The longest-lasting associations occur between mothers and their calves, but female bands also have relatively stable membership; sub-adult and adult males can also form lasting associations, often with other males from their natal pod.

Bottlenose dolphins often display forward or side breaches, somersaults, and tail slaps, and frequently bow-ride vessels. Mixed herds of bottlenose dolphins and pilot whales are commonly observed in offshore habitats, and the species associates also with white-beaked, Atlantic white-sided, common and Risso's dolphins, and sometimes with fin, sei, humpback and right whales. Bottlenose dolphins can be quite aggressive, not only to conspecifics but also towards smaller species. In the Moray Firth and Cardigan Bay, harbour porpoises have been killed in attacks from bottlenose dolphins (Ross and Wilson 1996; Jepson and Baker 1998).

The bottlenose dolphin is a very catholic feeder, taking a wide variety of benthic and pelagic fish (both solitary and schooling species), as well as cephalopods and shellfish. Haddock, saithe, cod, hake, blue whiting, snipefish, hake, mullet, silvery pout, eels, salmon, trout, bass, sprat and sandeels, as well as octopus and other cephalopods have all been recorded in the diet of European animals (Desportes 1985; Evans 1987; Relini *et al.* 1994a; Salomón *et al.* 1997; Silva and Sequeira 1997; Santos Vázquez 1998). Individual animals may feed solitarily, but the species also herds fish co-operatively, trapping them against the water surface, shore line, or tidal interface (Liret *et al.* 1994; Wilson *et al.* 1997).

Global distribution

The bottlenose dolphin has a worldwide distribution in tropical and temperate seas of both the southern and northern hemispheres. It occurs in all oceans and in a diverse range of habitats, from shallow estuaries and bays to the continental shelf edge and beyond into deep open oceans. In coastal waters, bottlenose dolphins often favour river estuaries, headlands or sandbanks, where there is uneven bottom relief and/or strong tidal currents (Lewis and Evans 1993; Liret *et al.* 1994; Wilson *et al.* 1997).

North Atlantic status

In the western North Atlantic, bottlenose dolphins occur inshore off New England during the summer months, but further north, as far as Nova Scotia, the distribution appears to be more offshore. In the eastern North Atlantic, they have been reported as far north as northern Norway and Iceland (Wells and Scott 1999).

Although overall population estimates do not exist, photo-identification studies indicate a resident population of around 130 bottlenose dolphins in the Moray Firth (Wilson *et al.* 1997), while the population in Cardigan Bay has been variously estimated at 130-350 individuals (Lewis 1992, Arnold *et al.* 1997). Neither population is closed, and recognisable individuals from each have been reported outside those regions. The inner Moray Firth population may be declining at an annual rate of over 5% (Sanders-Reed *et al.* 1999). In western Ireland, a cumulative minimum estimate of 115 dolphins inhabit the Shannon Estuary (Ingram *et al.* 1999), and a collaborative photo-identification project has catalogued 85 individuals in the Channel, including north-west France (Liret *et al.* 1998). Animals off the French coast form very stable groups that are resident in small areas, whereas those along the southern English coast are wider-ranging, and possibly make seasonal movements between Cornwall and Sussex (Evans 1992; Williams *et al.* 1996; Liret *et al.* 1998; Wood 1998).

NW European distribution

Along the Atlantic seaboard of Europe, bottlenose dolphins are locally common near-shore off the coasts of Spain, Portugal, north-west France, western Ireland, the Irish Sea (particularly Cardigan Bay), and north-east Scotland (especially the Moray Firth), with smaller numbers in the Channel, particularly in the western portion from the Channel Islands west. The species also occurs around the Faroe Islands (Bloch 1998).

Largest numbers have been observed off western Ireland and, in the vicinity of the shelf break, south-west of Ireland south towards the French coast. However, the species also occurs further offshore in deep waters of the North Atlantic, often in association with long-finned pilot whales and Atlantic white-sided dolphins (Evans 1991; Pollock *et al.* 2000). Sightings of groups have been made in the vicinity of the Rockall Bank and over the Wyville Thompson Ridge and Ymir Ridge.

The greatest numbers at most UK sites have been recorded between July and October (with a secondary peak in some localities in March-April), although some animals are present near-shore in every month of the year (Evans 1992; Wilson *et al.* 1997). Williams *et al.* (1996) found some evidence of seasonal movements along the south coast of England: during winter most sightings occurred around Cornwall; during spring, most were eastwards as far as the east Sussex coast; by summer, most sightings were from Lyme Bay eastwards; and during autumn, most sightings occurred off the Dorset coast east to the Isle of Wight. In the Faroes, the species is seen mainly in March and between July and October (Bloch 1998).

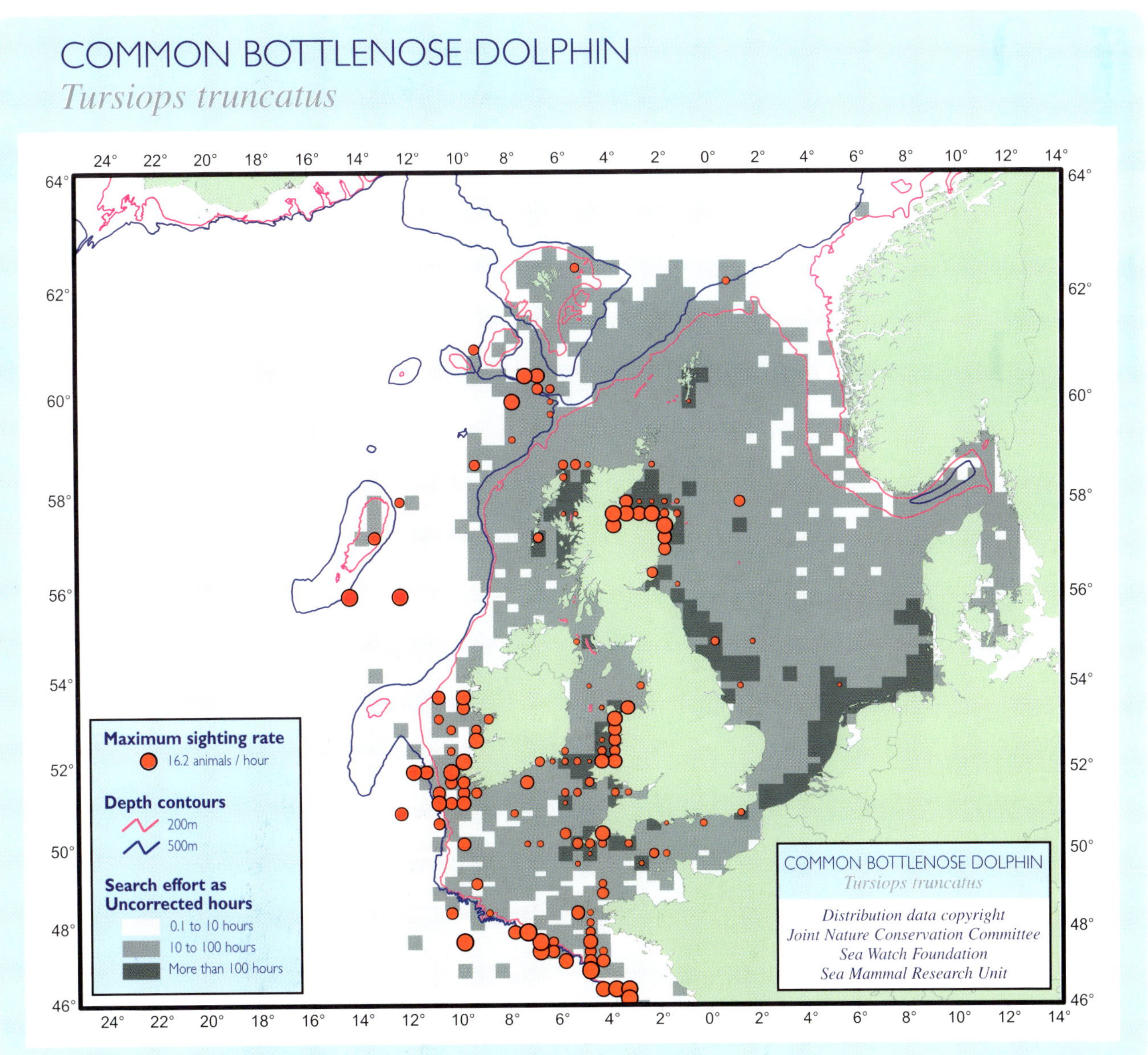

STRIPED DOLPHIN
Stenella coeruleoalba

The striped dolphin is a small, slender dolphin, about 2.0-2.4 m in length, superficially resembling the common dolphin. It has a tapering forehead with a distinct groove separating a black beak of medium length and like the common dolphin, a dark back and lighter flanks.
The centrally-placed slender dorsal fin is sickle-shaped or nearly erect. The species is distinguished by a white or light grey blaze that becomes V-shaped from above and behind the eye, one line narrowing to a point below the fin, and the lower one extending towards the tail.
There are also two black lines on the flanks, one extending from near the eye to the anus, and a second from the eye to the flippers.

Striped dolphins are very sociable animals, occurring in groups of hundreds or even thousands of individuals, although in European waters, group sizes of 6-60 are most common. In British and Irish waters, however, most sightings are of single animals or small groups numbering less than 10, often in mixed schools with common dolphins. Groups can show strong segregation by age (for example, entirely immature schools), with some evidence of the sexes being segregated outwith the breeding season (70% of recent strandings in the UK and Ireland have comprised males - Collet and Evans, in press).

A fast swimmer, the striped dolphin frequently breaches clear of the water, and may bow-ride, although the species does not appear to be attracted to vessels to the same extent as common dolphins.

In the eastern North Atlantic, the striped dolphin feeds on a variety of mesopelagic and benthic fish, including sprat, blue whiting, *Trisopterus* spp., silvery pout, whiting, hake, scad, bogue, anchovy, and gobies. Squid are also frequently taken, as are crustaceans (Desportes 1985; Berrow and Rogan 1995; Santos Vázquez 1998).

Visual and acoustic surveys in the western Mediterranean indicate that striped dolphins may concentrate to feed along the shelf edge (500-1,000 m depth) at night, and then move offshore during the day before returning to shallower waters after dusk (Gannier and David 1997).

Global distribution

The striped dolphin has a worldwide distribution, occurring in both the southern and northern hemispheres mainly in tropical, subtropical and warm-temperate oceanic waters. It tends to occur well beyond the continental shelf in depths of 1,000 m or deeper, but it occasionally comes over the shelf and has been recorded in waters of 60 m depth or less (Forcada *et al.* 1990).

North Atlantic status

In the North Atlantic, the striped dolphin typically occurs from Virginia south to Florida in the west (with extra-limital records from southern Greenland and the Gulf of Mexico), and the British Isles south to Spain and Portugal and throughout the Mediterranean in the east (but with records also from the Faroe Islands, Iceland and Norway, central North Atlantic and the Canaries) (Perrin *et al.* 1994; Isaksen and Syversen 2002).

The only striped dolphin population estimate for the eastern North Atlantic is 73,843 individuals (CI: 36,113-150,990) for an area of the continental shelf extending south-west of Ireland to France and north-west Spain (excluding the Bay of Biscay), and westwards to 20° W (Goujon *et al.* 1994).

Approaches and off southern Ireland; however, occasional sightings and strandings have occurred as far north as Shetland, and the species has been seen in deep waters off the continental shelf north to 62° N (Evans 1992). Most near-shore records from the UK have been between July and December (Evans 1992; Collet and Evans, in press).

NW European distribution

In the eastern North Atlantic, striped dolphins occur mainly offshore west of the Iberian Peninsula and France, and in the Bay of Biscay (Forcada *et al.* 1990; Perrin *et al.* 1994; Coles *et al.* 2001).

In UK waters, the species is rare. With its normal distribution reaching its northern limits at 50° N, most casual records come from the South-west Channel

STRIPED DOLPHIN
Stenella coeruleoalba

24° 22° 20° 18° 16° 14° 12° 10° 8° 6° 4° 2° 0° 2° 4° 6° 8° 10° 12° 14°

64° 62° 60° 58° 56° 54° 52° 50° 48° 46°

Maximum sighting rate
89 animals / hour

Depth contours
200m
500m

Search effort as Uncorrected hours
0.1 to 10 hours
10 to 100 hours
More than 100 hours

STRIPED DOLPHIN
Stenella coeruleoalba

Distribution data copyright
Joint Nature Conservation Committee
Sea Watch Foundation
Sea Mammal Research Unit

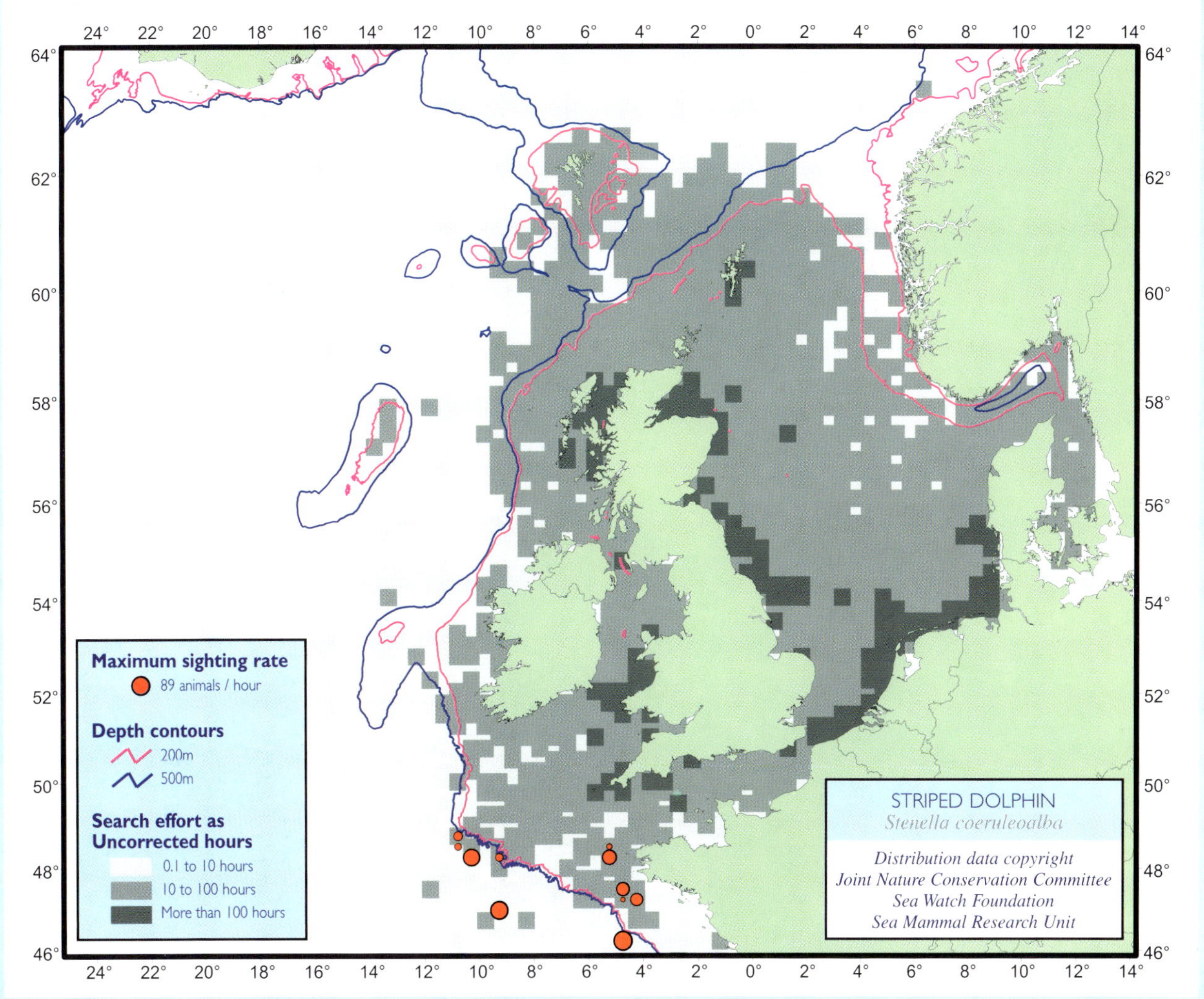

FRASER'S DOLPHIN

Lagenodelphis hosei

Fraser's dolphin was not recognised to science until as recently as 1956, when it was described from a skull picked up in the nineteenth century on a beach in Sarawak (Fraser 1956). It is a small, stocky dolphin with a very short but distinct beak, small flippers and dorsal fin. In general shape it is not as slender as the common dolphin but not as stocky as white-beaked or white-sided dolphins. It has a black back, grey flanks, and a white to pink belly with broad darker stripes from the beak to the flipper and from the eye to the genital area. Males typically grow to 2.4 m, females to 2.2 m (Perrin et al. *1994).*

It is a gregarious species, most schools being composed of 100-1,000 individuals, although small groups of 4-15 have also been sighted (Perrin *et al.* 1994; Collet *et al.* in press). The species' social structure is unknown, although groups of mixed sexes and ages have been reported. It is a fast, very active swimmer, attaining speeds of 20 km/hr and more (Collet *et al.* in press). It commonly breaches, and approaches boats to bow-ride. It often associates with other species such as melon-headed whale, false killer whale, spinner, spotted and striped dolphins, and sometimes sperm whale.

The species apparently feeds on mesopelagic fishes, shrimps and squids (Robinson and Craddock 1983; Ross 1984). Animals stranded in northern France had been feeding on blue whiting, *Trisopterus* spp. whiting, and cuttlefish (Van Bree *et al.* 1986).

Global distribution and North Atlantic status

Although poorly known, Fraser's dolphin appears to have a worldwide distribution in tropical pelagic waters, occasionally entering temperate coastal areas (South Australia, France, and Scotland). It is found mainly in equatorial seas around areas of upwelling such as the deep waters around oceanic islands, where the species sometimes comes very close to the coast (Collet *et al.* in press).

Live sightings in the North Atlantic come from the Caribbean, Azores and Canaries. No information exists on population status.

NW European distribution

The first record of the species in European waters was a mass stranding on the north Brittany coast of France in May 1984 (Van Bree *et al.* 1986). There has been only one confirmed record in British waters; a single male stranded on South Uist (West Scotland) in 1996 (Bones *et al.* 1998).

SHORT-BEAKED COMMON DOLPHIN

Delphinus delphis

The short-beaked common dolphin is usually easy to identify at sea by the distinctive light-coloured hour-glass pattern on the lower flanks (tan or yellowish anterior to the fin, and light grey in the posterior portion). The rest of the upper body is dark but rather variable in coloration. Often the best field character, however, is the dark triangle below the dorsal fin. It is a small, slender and swift dolphin with a pronounced beak. Females grow to a length of* c. *2.1 m, males to* c. *2.4 m.

Although social structure in this species is poorly known, common dolphins are gregarious animals; average group sizes observed in north-west European waters are between six and ten Goujon 1996), though large schools of dozens or even hundreds have frequently been recorded. Average group size in the present data-set is 14 individuals. Common dolphins frequently breach and often bow-ride.

The diet of common dolphins comprises a wide range of small fish and squid. There are relatively few published accounts of analyses of stomach contents in the north-east Atlantic. Where they occur, mackerel, sprat, pilchard, anchovy, scad, hake, blue whiting, *Trisopterus* spp. and cephalopods, including the lesser flying squid (*Todaropsis eblanae*), are known to be exploited (Collet 1981; Pascoe 1986; Santos Vázquez 1998; Silva 1999). Generally, the prey species most frequently consumed in the north-east Atlantic appear to be pelagic, schooling fishes.

Global Distribution

Common dolphins are among the most abundant cetaceans throughout the world's warm-temperate and tropical waters. Until recently, only one species was recognised in this genus, but recently, at least three species have been described in this cosmopolitan group. In the temperate North Atlantic all individuals appear to be *D. delphis*, the short-beaked or offshore common dolphin, which reportedly favours deep water habitats (Evans, 1994). The long-beaked common dolphin *D. capensis* has a disjunct distribution around the world, mainly in warm temperate and tropical coastal waters..

North Atlantic status

The short-beaked common dolphin is the most numerous offshore cetacean species in the temperate north-east Atlantic. Two population estimates have been made in separate but overlapping areas to the south-west of Britain.

The SCANS survey in July 1994 (Hammond *et al.* 1995) covered the Celtic Sea to approximately 11° W and 48° S, and resulted in an estimate of 75,500 animals in that area (95% CI: 23,000-249,000).

The MICA survey in July and August of 1993 covered a more southerly and westerly region, extending as far south as 43° N and as far west as 19°-21° W. The number of common dolphins in this area was estimated at 62,000 (95% CI: 35,000-108,000; Goujon 1996). There was a small amount of overlap in the areas covered by the two surveys, and Goujon (1996) suggested a total population of around 120,000 common dolphins in the two areas combined.

NW European distribution

The common dolphin in European waters is mainly distributed south of around 60° N, in Atlantic waters. Off the western coasts of Britain and Ireland, the species is found in continental shelf waters, notably in the Celtic Sea and Western Approaches to the Channel, and off southern and western Ireland.

It has frequently been seen in the Sea of Hebrides in summer. It has been observed occasionally in the North Sea, also mainly in summer (June to September). The limited survey effort achieved in deeper Atlantic waters suggests that common dolphins are widespread in waters well offshore to the west of Ireland.

SHORT-BEAKED COMMON DOLPHIN

Delphinus delphis

Maximum sighting rate

126 animals / standard hour

Depth contours

200m

500m

Search effort as Standardised hours

0.1 to 10 hours

10 to 100 hours

More than 100 hours

SHORT-BEAKED COMMON DOLPHIN

Delphinus delphis

Distribution data copyright
Joint Nature Conservation Committee
Sea Watch Foundation
Sea Mammal Research Unit

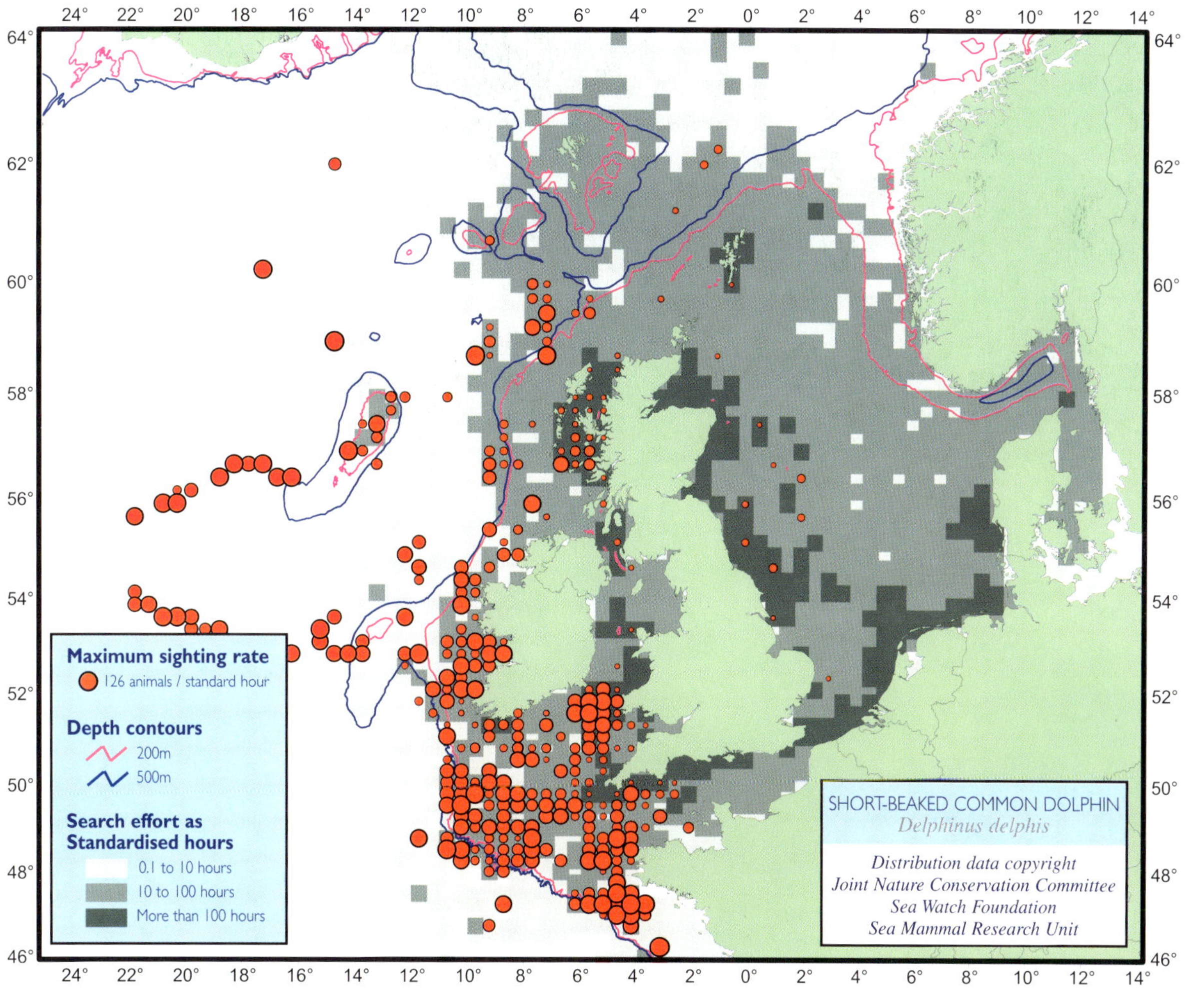

WHITE-BEAKED DOLPHIN

Lagenorhynchus albirostris

The white-beaked dolphin is a stout dolphin, about 2.5-2.7 m in length; it has a short, often white, beak. The back is black, except for a pale grey to whitish area behind the dorsal fin that extends from the grey to whitish blaze on the flanks. This forms a pale 'saddle', a diagnostic feature distinguishing this species from the Atlantic white-sided dolphin.

Although usually found in smaller groups than the Atlantic white-sided dolphin, mostly numbering less than 10 individuals, herds of up to 50 are not uncommon, and aggregations can comprise 100-500 animals in northern parts of their range and also in the North Sea (Evans 1991; Kinze *et al.* 1997). There appears to be segregation by age and sex, with juvenile groups sometimes distinct from groups of adults with calves (Reeves *et al.* 1999b). The species often displays forward, vertical or side breaches, and frequently bow-rides vessels. Mixed herds of white-beaked dolphins have been observed with Atlantic white-sided, bottlenose, common and Risso's dolphins, and sometimes also with fin, sei, and humpback whales, long-finned pilot whales, and killer whales. In common with other dolphin species, co-operative food herding has frequently been observed (Evans 1987). Though not as agile as common or striped dolphins, white-beaked dolphins are relatively fast swimmers, usually travelling at speeds of *c.* 6-12 km/hr, although they can attain bursts of speed of 30 km/hr (Evans and Smeenk, in press).

White-beaked dolphin diet consists of a variety of fish including mackerel, herring, cod, capelin, whiting, haddock, *Trisopterus* spp., navaga, hake, scad, snow crab, and various species of sandeels, gobies, flatfishes, and scaldfishes; and amongst cephalopods, the octopus *Eledone cirrhosa* (Santos Vázquez *et al.* 1994; Kinze *et al.* 1997; Reeves *et al.* 1999b). Analyses of stomach contents from various parts of the North Sea and from Newfoundland have revealed cod, whiting and hake as predominant prey (Santos Vázquez *et al.* 1994; Kinze *et al.* 1997).

Global distribution

The white-beaked dolphin is restricted to temperate and sub-Arctic seas of the North Atlantic. It is usually found over the continental shelf in waters of 50-100 m depth.

North Atlantic status

In the North Atlantic, the white-beaked dolphin ranges from central west Greenland to Cape Cod in the west, and western Svalbard and Novaya Zemlya to the French coast in the east, with extra-limital records in the Bay of Biscay, the coasts of Spain and Portugal, and approaches to the Mediterranean (Reeves *et al.* 1999b).

SCANS line transect surveys in July 1994 indicated a population estimate of 7,856 white-beaked dolphins (95% CI: 4,032-13,301) for the North Sea and Channel (Hammond *et al.* 1995). The estimate for all *Lagenorhynchus* dolphins (including those of unidentified species) was 11,760 (95% CI: 5,867-18,528).

NW European distribution

The white-beaked dolphin occurs over a large part of the northern European continental shelf. It is recorded more frequently in the western sector of the central and northern North Sea across to western Scotland and south to western Ireland; it also occurs occasionally off southern Ireland, in the Irish Sea and western Channel, and even to the northern Bay of Biscay (Pollock *et al.* 1997, 2000; Coles *et al.* 2001).

Although present over the continental shelf year-round in near-shore UK waters, the species has been observed most frequently between June and October (Evans 1992; Northridge *et al.* 1995).

WHITE-BEAKED DOLPHIN
Lagenorhynchus albirostris

Maximum sighting rate
7.5 animals / standard hour

Depth contours
200m
500m

Search effort as Standardised hours
0.1 to 10 hours
10 to 100 hours
More than 100 hours

WHITE-BEAKED DOLPHIN
Lagenorhynchus albirostris

Distribution data copyright
Joint Nature Conservation Committee
Sea Watch Foundation
Sea Mammal Research Unit

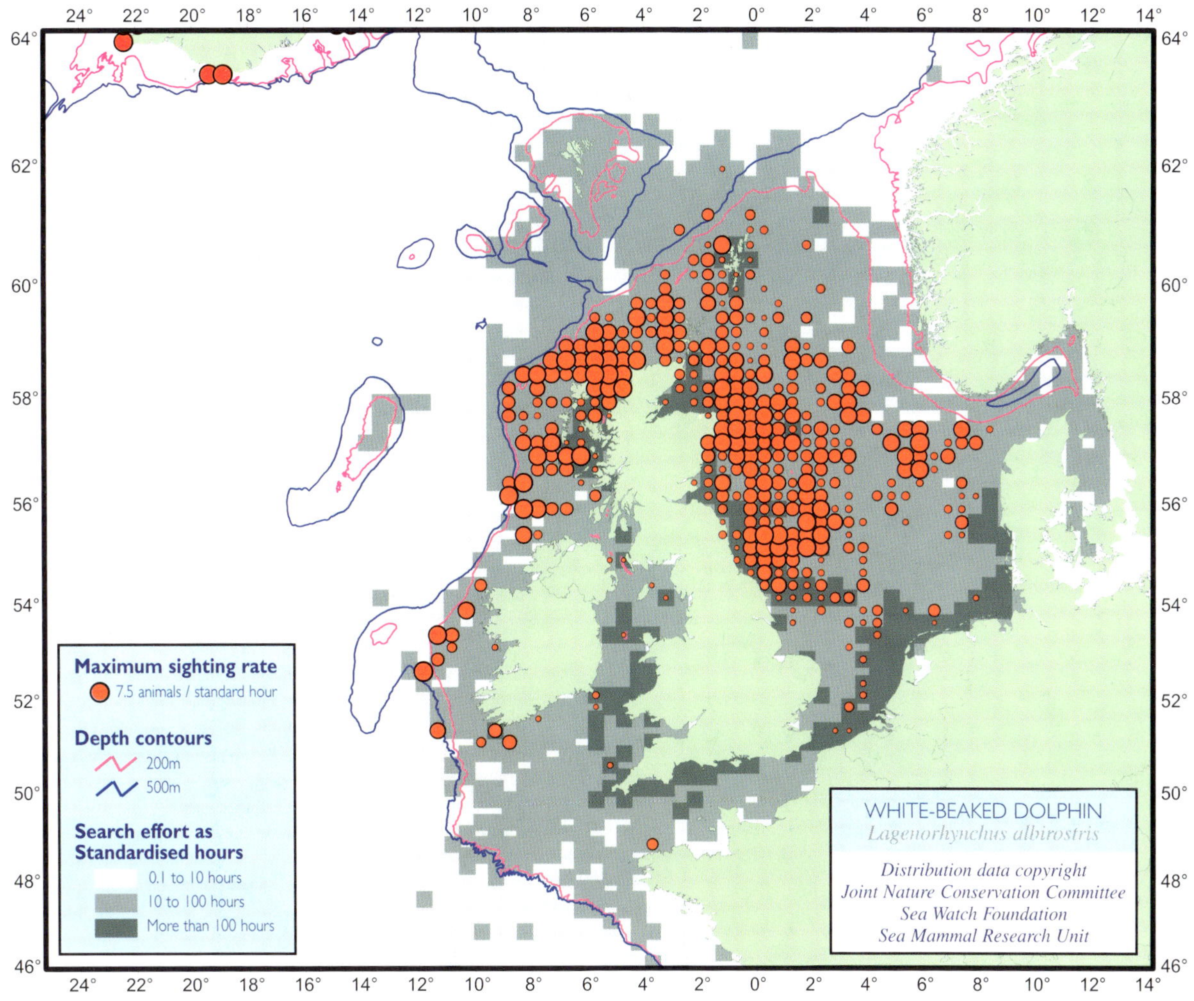

ATLANTIC WHITE-SIDED DOLPHIN

Lagenorhynchus acutus

The Atlantic white-sided dolphin is superficially rather similar to the white-beaked dolphin. The two species may form mixed herds that are sometimes very large. A rather bulky dolphin, reaching sizes of nearly 3 m in length, the white-sided dolphin is best distinguished from the white-beaked dolphin by its all-black back and elongated yellow-ochre band on the side, extending backwards from the upper edge of a long, white, oval blaze. In common with the white-beaked dolphin, it breaches frequently, but is much less inclined to bow-ride vessels. An agile and fast swimmer, it can swim over long distances at average speeds of at least 14 km/hr (Mate and Stafford 1994), but may achieve much higher speeds. It associates with several other species of cetaceans, particularly white-beaked dolphins, sometimes bottlenose and common dolphins, pilot whales, fin whales, humpbacks and sperm whales.

The species is very gregarious, with observed group sizes frequently numbering in the tens to hundreds, sometimes up to 1,000, particularly offshore. Within large aggregations, subgroups of 2-15 animals can often be distinguished (Reeves *et al.* 1999a). Groups comprise both sexes and are of mixed age, although some age segregation has been suggested from two mass strandings (Sergeant *et al.* 1980; Rogan *et al.* 1997).

The diet of Atlantic white-sided dolphins consists of a wide variety of fish, particularly gadoids such as blue whiting, whiting, *Trisopterus* spp., cod and hake, clupeids, particularly herring, and silvery pout, lantern fishes, pearlsides, mackerel, horse mackerel and salmonids (Sergeant *et al.* 1980; Desportes 1985; Rogan *et al.* 1997; Couperus 1997, 1999; Evans and Smeenk, in press). The stomachs of 46 white-sided dolphins by-caught in the Dutch mid-water trawl fishery for mackerel and scad between 1992 and 1994 (mainly February-March) south-west of Ireland, revealed a predominance of mackerel, scad being notably absent. The mackerel were apparently taken near the surface and behind trawlers hauling in nets; fish otoliths of earlier meals indicated a diet dominated by silvery pout, lantern fishes and pearlsides (Couperus 1997). A variety of squid is also taken, mainly of the family Ommastrephidae (Rogan *et al.* 1997; Couperus 1997, 1999).

Global distribution

The Atlantic white-sided dolphin is confined to temperate and sub-Arctic seas of the North Atlantic. Its status in the central North Atlantic is poorly known, but it occurs abundantly on both sides. In the west, it ranges from the Davis Strait and Greenland south to the Gulf of St Lawrence and Cape Cod (occasionally south to Chesapeake Bay). In the east, it occurs from Iceland, southern Svalbard and the Barents Sea (certainly east to Murmansk, *c.* 35° E), south to the Bay of Biscay and occasionally to Portugal, the western Mediterranean and the Azores. It is generally less common over continental shelves than over slopes (not so in the NW Atlantic) and in deeper waters, and more abundant north of 56° N than south of this latitude, although large numbers have been recorded off south-west Ireland and into the Celtic Sea at 47° N (Leopold and Couperus 1995; Couperus 1997).

White-sided dolphins live mainly in cool waters (7-12° C), particularly seaward or along the edges of continental shelves at depths of 100-500 m, but they may also be numerous in much deeper, oceanic waters (Leopold and Couperus 1995; Pollock *et al.* 2000). They seem to favour areas of high bottom relief and around deep submarine canyons (Selzer and Payne 1988). Nevertheless, they sometimes come into continental shelf waters such as those in the north-western North Sea, and may enter fjords or inlets.

North Atlantic status

Population estimates are difficult to obtain because of confusion with the white-beaked dolphin (Hammond *et al.* 1995). This is not a problem in North American waters, however, where white-beaked dolphins are mainly distributed further offshore. Here, line-transect surveys in the Gulf of Maine, the Gulf of St Lawrence and the Labrador Sea (*c.* 37-48° N) yielded a combined population estimate of 27,200 individuals (CV=0.43) (Palka *et al.* 1997). Over the European continental shelf, white-sided dolphin sightings are usually greatly outnumbered by those of white-beaked dolphins (Evans 1992; Hammond *et al.* 1995; Northridge *et al.* 1997). Hammond *et al.* (1995) estimated 11,760 (95% CI: 5,867-18,528) *Lagenorhynchus* spp. in the North Sea, Celtic Sea and Baltic during July 1994, but most of these (7,856; 95% CI: 4,032-13,301) were white-beaked dolphins, and an unknown proportion of the remainder were not specifically identified.

NW European distribution

In north-west Europe, the distribution of the Atlantic white-sided dolphin is concentrated in a broad zone from west of Ireland to the north and north-west of Britain as well as north of Finnmark and north-west Russia. Numbers are rather fewer around western Ireland and in the South-west Approaches to the Channel and Celtic Sea, and also in the north-western North Sea and the Norwegian Sea (Evans 1991, 1992; Pollock *et al.* 1997, 2000). The species is rare in the Irish Sea, the Channel, the Southern and German Bights of the North Sea and the central and north-eastern North Sea. It is also rare in the Kattegat, Skagerrak and Belt Seas, although some groups have been recorded in these waters; there is only one record from the Baltic Sea (Kinze *et al.* 1997).

Little is known of seasonal movements of white-sided dolphins. They are found in deep waters around the north of Scotland throughout the year. They seem to enter the North Sea mainly in summer, and may also follow mackerel as they spawn off south-west Ireland in February/March (Couperus 1997), although this is not apparent from the Atlas data. Couperus (1997) has speculated that white-sided dolphins that have been observed at the shelf break off south-west Ireland in February/March must come from deeper, offshore Atlantic waters further west, rather than from further north, since the trawler fleet further north (where mackerel are also exploited) has not bycaught animals earlier in the year. This would suggest that white-sided dolphins may range much further into the open Atlantic than previously thought.

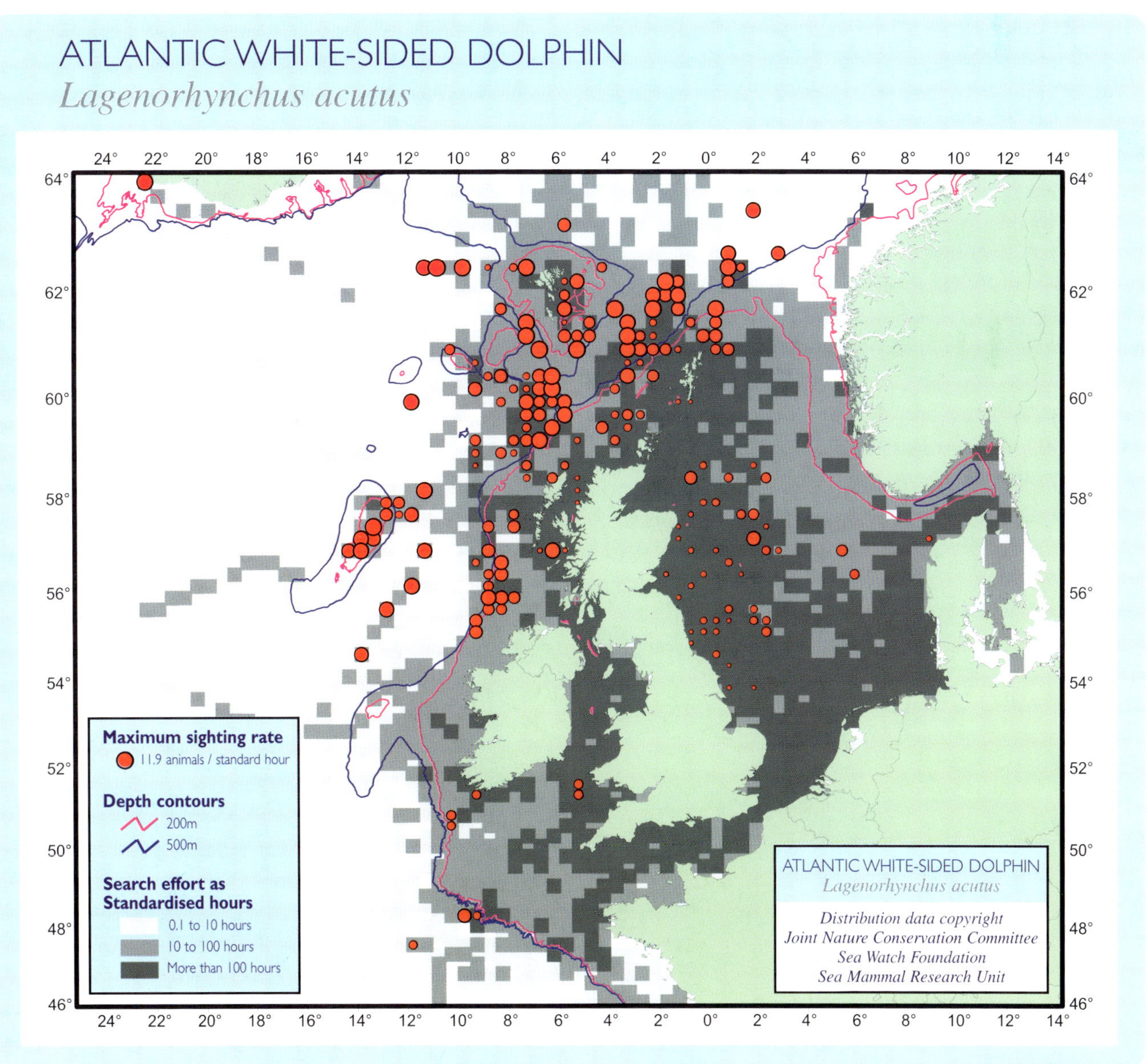

RISSO'S DOLPHIN

Grampus griseus

Risso's dolphin is a large, robust dolphin around 3.5 m in length, with a blunt, rounded head, and a slight melon but no beak. It is distinctively greyish in colour (whitening with age), often with numerous white scars on the flanks. It is a relatively slow swimmer (4-12 km/hr), and although usually slightly wary of vessels, it occasionally bow-rides (mainly juveniles), and regularly engages in a variety of surface behaviour including breaching (particularly juveniles), spyhopping, tail-slapping, and communal diving.

It is a gregarious species, forming small to medium-sized groups, typically ranging from 2-50 animals. In European waters, the modal group size is 6-12 off the UK, 15-20 off Spain. However, it has been recorded singly as well as in temporary aggregations of several hundreds or even thousands (Kruse *et al.* 1999; Evans, in press). In the North Atlantic, Risso's dolphins have occasionally been seen swimming with other cetaceans, including long-finned pilot whales, white-beaked and Atlantic white-sided dolphins, and bottlenose dolphins.

Risso's dolphins have been reported to feed mostly on cephalopods, although small fish are also taken (Kruse *et al.* 1999). Analysis of stomach contents of five individuals from British waters indicated a diet of mainly octopus *Eledone cirrhosa*, but also cuttlefish *Sepia officinalis*, sepiolids and small bottom-dwelling squid such as *Loligo forbesi* and *Todarodes sagittatus* (Clarke and Pascoe 1985; Zonfrillo *et al.* 1988; Santos Vázquez 1998).

Global distribution

Risso's dolphins occur in virtually all of the world's oceans between 60° S and 60° N, although there may be a gap in distribution in the tropical Atlantic (Kruse *et al.* 1999).

The species does not appear to be common anywhere, with the possible exception of some waters off California.

Primarily a warm water (ranging from 4.5-28° C) pelagic species (Baird and Stacey 1991), Risso's dolphin generally prefers continental slope waters. In the eastern Pacific, the species typically occurs seaward of the 180 m depth contour and inhabits coastal areas only where the continental shelf is near the shore (Leatherwood *et al.* 1980; Kruse 1989). The average depth of water in which sightings have been made in this region is 1,000 m. Hain *et al.* (1981) and Kenney and Winn (1986, 1987) also recorded frequent use of the continental shelf edge off eastern USA. In UK continental shelf seas, Risso's dolphins have been recorded mainly over slopes of 50-100 m depth (Evans, in press). By contrast, in the Mediterranean Sea, the species has been noted mainly at between 500 and 1,000 m depths (Fabbri *et al.* 1992; Gannier and Gannier 1994; Cañadas and Sagarminaga 1996).

North Atlantic status

As a comparatively uncommon species there have been no attempts to estimate Risso's dolphin abundance over wide areas in the north-east Atlantic. Waring *et al.* (1999) report the results of four surveys off eastern North America. The most complete of these indicated a population estimate of 16,818 (CV=0.52). Atkinson *et al.* (1999) identified at least 142 individuals over two summers in the north-western Minch off western Scotland.

NW European distribution

In north-west Europe, Risso's dolphin appears to be a continental shelf species.. Most sightings are from western Scotland, with the waters surrounding the Outer Hebrides forming an obvious centre of distribution. There are other clusters of sightings in the southern Irish Sea and off south-west Ireland. There are few records from the central and southern North Sea and the Channel (although there are a number of casual sightings in the western portion south towards the Bay of Biscay - Evans 1992; Coles *et al.* 2001). Sightings around Shetland and Orkney, and off Norway are amongst the furthest north in the eastern Atlantic (Øien 1987; Evans 1996; Pollock *et al.* 2000). A few records come from waters immediately over the shelf break, but none from deeper areas.

There appears to be some seasonality in patterns of occurrence, with Risso's dolphins being seen more frequently offshore near the continental shelf edge in winter (October to May). All records near the Wyville Thompson Ridge were made from October to December, those on the shelf edge south-west of Ireland during May (there was very little effort in this region from October to April). By contrast, the highest sightings rates in the Minch were between May and September. Most sightings from the Irish Sea were also between July and September. Near-shore records off south-west Ireland were obtained primarily between May and August, and in the northern North Sea in July and August, although some animals were present off north-east Scotland and Shetland in winter.

RISSO'S DOLPHIN
Grampus griseus

Maximum sighting rate
1.8 animals / standard hour

Depth contours
200m
500m

Search effort as Standardised hours
0.1 to 10 hours
10 to 100 hours
More than 100 hours

RISSO'S DOLPHIN
Grampus griseus

Distribution data copyright
Joint Nature Conservation Committee
Sea Watch Foundation
Sea Mammal Research Unit

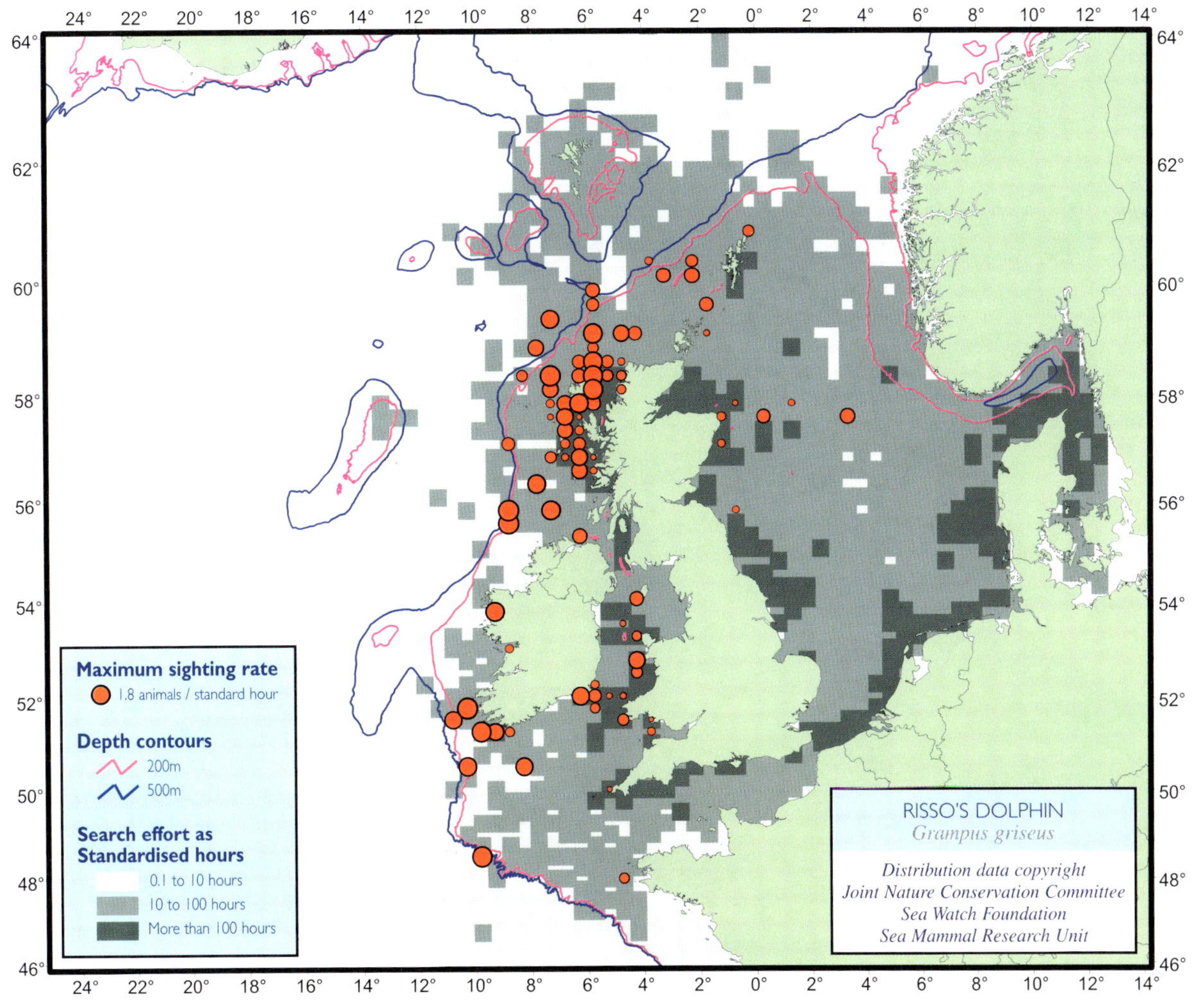

MELON-HEADED WHALE

Peponocephala electra

The melon-headed whale is a medium-sized dolphin about 2.5 m in length. It has a slender, almost black, torpedo-shaped body with a centrally placed, sickle-shaped dorsal fin, a triangular-shaped head and a rounded forehead. The slightly underslung jaw presents a very indistinct beak, often with white lips. The flippers have pointed tips.

An apparently gregarious species, the melon-headed whale is usually observed in large herds ranging from 50 to 1,500 individuals (Bryden *et al.* 1977; Perryman *et al.* 1994). Mass strandings have been reported on several occasions, often showing a sex ratio in groups of two females to every male. The species is a fast swimmer, often breaking the surface as a tightly packed group; it commonly bow-rides vessels,. It is known to breach and spyhop. Melon-headed whales have been observed associating with other cetaceans, particularly Fraser's dolphin (Perryman *et al.* 1994). The varied diet includes fish, ommastrephid squid and shrimps (Perryman *et al.* 1994). The species has also been reported herding and possibly attacking small dolphins (*Stenella* spp.) escaping from tuna seine nets in the tropical Pacific (Leatherwood and Reeves 1983).

Global distribution and North Atlantic status

Melon-headed whales appear to have a worldwide distribution, occurring in deep tropical and subtropical seas mainly between 40° N and 35° S. They are usually observed seaward of the edge of continental shelves, and around oceanic islands.

There is no information on the abundance of melon-headed whales in the North Atlantic, although large herds numbering a few hundreds have been seen in the eastern Caribbean (Evans, unpublished data).

NW European distribution

The only record from Europe of this species is of a skull found near Charlestown, Cornwall, found in September 1949, and originally misidentified as white-beaked dolphin (Mikkelsen and Sheldrick 1992).

FALSE KILLER WHALE

Pseudorca crassidens

False killer whales are 5-6 m in length and have a slender, almost all-black torpedo shaped body with a tall, usually sickle-shaped dorsal fin slightly behind the middle of the back. The head is small and narrow, tapering to overhang the lower jaw. An area of light grey is often present on the sides of the head and there can also be a blaze of grey on the chest between the flippers. The flippers are long, narrow and tapered, with a distinctive broad hump on the front margin near the middle.

The species appears to be highly social. Although groups of 10-50 animals are typical, larger herds numbering 600-800 have been reported (Ross 1984; Leatherwood *et al.* 1988). Occasional mass strandings of false killer whales may comprise animals of mixed age and sex groups (Sergeant 1982; Odell and McClune 1999). Fast-swimming and active, they can breach clear of the water, and sometimes approach and bow-ride vessels. They commonly associate with other cetaceans, such as bottlenose dolphins.

The diet of false killer whales is very varied, and includes many species of squid and large fish (Ross 1984; Baird *et al.* 1989). The species is also known to prey on small dolphins (e.g. *Stenella*, *Delphinus*) in the purse seine fishery for tuna in the tropical eastern Pacific, and has been recorded attacking sperm whales and on one occasion a humpback whale calf (Hoyt 1983; Odell and McClune 1999).

Global distribution and North Atlantic status

The false killer whale is a pelagic species with a worldwide distribution, occurring mainly in deep tropical to warm temperate waters, usually seaward of continental shelf breaks. It has also been observed around oceanic islands and in waters of 200 m depth or less (Kasuya 1971; Odell and McClune 1999).

In the north-east Atlantic, the species has been recorded only occasionally north of the British Isles, most reports having come from the Bay of Biscay south to the Canary Islands.

NW European distribution

In Britain, there have been only a few strandings, all involving large groups: *c.* 150 in the Dornoch Firth, north-east Scotland in October 1927; *c.* 25 along the Carmarthen and Glamorgan coasts, south Wales, in May 1934; and *c.* 75 along the east coast of Britain in 1935 (Fraser 1934, 1946). There have been no strandings since then. Strandings have also occurred in Holland and Denmark.

There have been five sightings of false killer whales in UK waters since 1976, at distances ranging from 5-54 km from land: two in the Atlantic off western Scotland, two in the northern North Sea off north-east Scotland, and one in the South-west Approaches south of Cornwall, (Evans1992). All sightings were made between July and November.

KILLER WHALE

Orcinus orca

Adult male killer whales may reach a length of 9 m, about 25% larger than adult females, and have a very tall, triangular and erect dorsal fin, which is sometimes tilted forwards. Immatures and adult females both have a smaller, sickle-shaped dorsal fin and cannot readily be distinguished from one another. When the animal surfaces, the grey saddle shows up over the black back, behind the dorsal fin. It has a conical-shaped black head, with a distinctive white oval patch above and behind the eye, an indistinct beak, white throat and large paddle-shaped flippers.

Killer whales live in very stable, matriarchal, extended family groups. Mothers form long-term, close associations with their sons, although both sexes may remain in their natal group throughout adulthood (Balcomb *et al.* 1982; Bigg *et al.* 1990). Mating with unrelated individuals probably occurs during brief periods of pod coalition, when members of different pods come together and exhibit much sexual activity (Osborne 1986; Heimlich-Boran 1988). A wide variety of behaviour has been observed, including various types of breaching, fluke- and flipper-slapping, lobtailing, and spyhopping (Jacobsen 1986). Non-predatory associations with other cetaceans (for example, minke whale and various pelagic dolphin species) have frequently been recorded (Jefferson *et al.* 1991). Most sightings in UK waters are of singles or groups of less than eight individuals (mean = 4.6), although groups of up to one hundred have been observed (Evans 1988; Pollock *et al.* 2000).

This species has one of the most varied diets of all cetaceans, ranging from fish and squid to birds, turtles, seals and other cetaceans (Hoyt 1990; Jefferson *et al.* 1991). Fish species taken in the eastern North Atlantic include herring, mackerel, cod, salmon, halibut, and bonito (Evans 1980; Couperus 1993; Ugarte and Similä 1993). Several studies in the Pacific suggest that transient pods feed primarily upon marine mammals, and resident whales mainly on fish (Felleman *et al.* 1991; Baird *et al.* 1992). Killer whales often feed co-operatively when hunting, particularly when pursuing marine mammal prey (Jacobsen 1986; Hoelzel 1991). When feeding, they may lunge at the surface, or engage in 'carousel' feeding, where schools of fish (herring) are herded into a tight ball, aided by tail slaps used to stun the fish (Ugarte and Similä 1993).

Global distribution

The killer whale has a worldwide distribution in tropical, temperate and polar seas in both the southern and the northern hemisphere, occurring at greatest abundance in colder waters at higher latitudes (Dahlheim and Heyning 1999). It is usually found within 800 km of continents (Perrin 1982) and although it generally prefers deep waters it occurs also in shallow bays, inland seas, and estuaries.

North Atlantic status

Although killer whale numbers in the North Atlantic appear to be greatest in sub-Arctic and Arctic waters, the distribution of the species extends south to the Caribbean, Azores, Madeira, Canaries and occasionally the western Mediterranean. No overall population estimates exist, but recent sightings surveys in the eastern North Atlantic, mainly between Iceland and the Faroe Islands, indicate a population in the region of somewhere between 3,500 and 12,500 animals (Gunnlaugsson and Sigurjónsson 1990).

NW European distribution

Killer whales are widely distributed in the deep North Atlantic and in coastal northern European waters, particularly around Iceland, the Faroe Islands and western Norway. In UK waters, it is most common off northern and western Scotland, but occurs also west and south of Ireland; it is rare in the central and southern North Sea, Irish Sea, and Channel.

In UK waters, killer whales occur in all months of the year, but have been recorded near-shore mainly between April and October (Evans 1988, 1992). Between Shetland and Norway, however, the species has been recorded regularly from November to March, commonly associating with purse seine fishing boats (Couperus 1993), and even taking mackerel out of the nets (Sea Watch, unpubl. data). Recent surveys north and west of Scotland suggest that killer whales concentrate along the continental slope north of Shetland during May and June (Bloor *et al.* 1996; Pollock *et al.* 2000), although these are the months when killer whales usually first appear in coastal waters around the Northern Isles and Outer Hebrides (Evans 1988). Seasonal movements may be associated with the distribution of particular prey; for example, seals are preyed upon close to land particularly from June to October when they haul out to breed, and pelagic fish such as herring are exploited off Møre (Norway) in February-March and around the Lofoten Islands in April-May and October-January (Similä 1997).

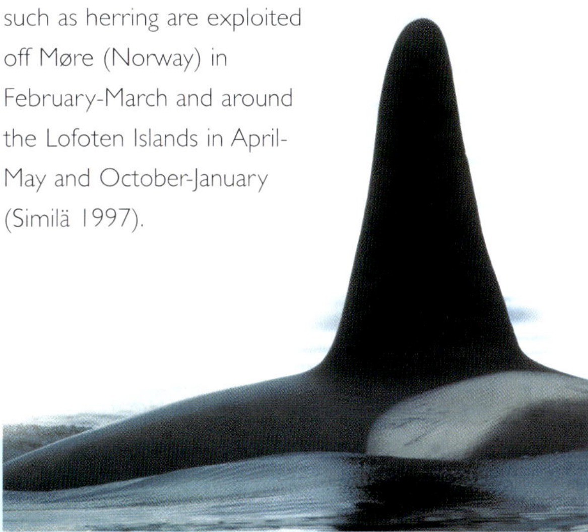

KILLER WHALE
Orcinus orca

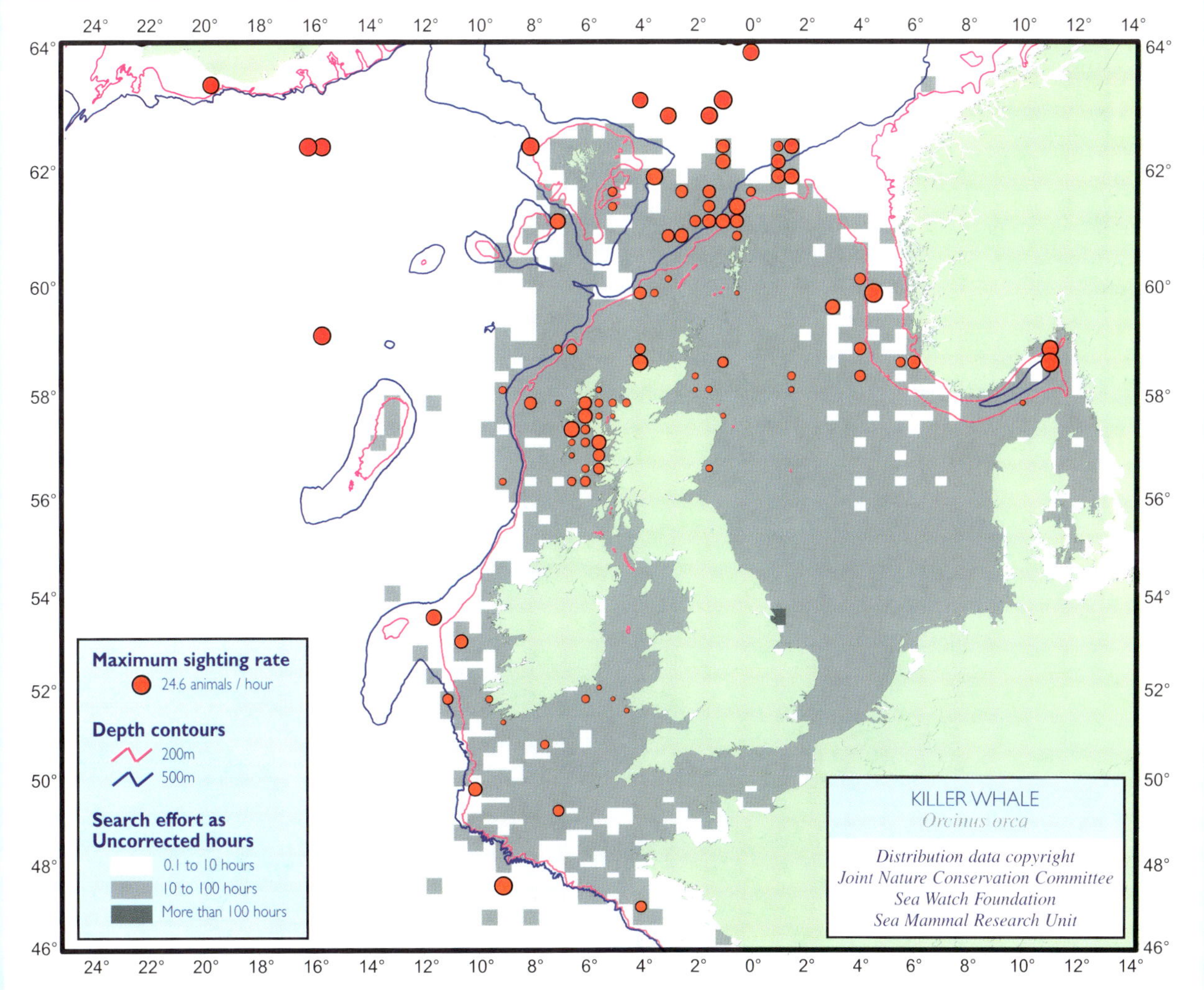

24° 22° 20° 18° 16° 14° 12° 10° 8° 6° 4° 2° 0° 2° 4° 6° 8° 10° 12° 14°
64° 62° 60° 58° 56° 54° 52° 50° 48° 46°
Maximum sighting rate
24.6 animals / hour
Depth contours
200m
500m
Search effort as Uncorrected hours
0.1 to 10 hours
10 to 100 hours
More than 100 hours
KILLER WHALE
Orcinus orca
Distribution data copyright
Joint Nature Conservation Committee
Sea Watch Foundation
Sea Mammal Research Unit

The species occurs mainly in deep waters (200-3,000 m) seaward and along the edges of continental shelves where bottom relief is greatest, although it may occasionally venture into coastal waters such as fjords and bays.

North Atlantic status

Wide-area transect surveys were made in the northern and north-eastern North Atlantic in the summers of 1987 and 1989. These indicated that few pilot whales occur north of Iceland (66° N) and that their core range is deep water south-west of the Faroes and south and west of Iceland. Surveys covered a larger proportion of the range in 1989 and the best estimate of numbers was 778,000 (CV=0.30; Buckland *et al.* 1993). Further west, Waring *et al.* (1999) reported their best estimate of numbers in the species' habitat between Virginia and the Gulf of St Lawrence to be 8,176 (CV=0.65). Given the gaps in these surveys, and the difficulties of estimating some parameters such as group size from ship-based surveys, an estimate of the total North Atlantic population cannot be made. The species occurs also in the Bay of Biscay south to the Iberian Peninsula, and is common in the Mediterranean Sea.

NW European distribution

The distribution map of the pilot whale highlights its deep water habitat, the species occurring in greatest numbers to the north of Scotland and south-east of the Faroes as well as along the shelf edge from southern Ireland south to the Bay of Biscay. Most records are from waters deeper than 200 m, with relatively few occurrences in shallower water around northern Scotland, the northern North Sea and the Channel. Occasionally, the species ventures into coastal waters in areas such as the Faroes, northern Scotland, western Ireland and the South-west Approaches to the

Channel, even entering fjords and bays. North of Scotland, highest sightings rates occurred over deeper areas (500-2,000 m) to the north and south of the Wyville Thompson Ridge although several sightings were from the western side of the Norwegian Rinne (along the 200 m depth contour). There are no records from the Porcupine Bank or its fringes. To the south-west of Ireland south to the Bay of Biscay, sightings are mostly along the continental shelf edge or beyond at depths of 200-2,000 m, occurring in all months with adequate survey effort.

There appears to be little seasonality in the pattern of sightings, although Evans (1980) found that incidental sightings peaked in the south-west English Channel and North Sea between November and January, when pods were frequently seen near vessels fishing for mackerel and when bycatches (in the Channel) were reported. Median group size ranged from 10-15 (maximum 200) between May and August, whereas for six out of eight months between September and April, it varied between 20 and 25 (maximum 1,000 - Evans 1992).

The species probably dives to a few hundred metres and its spatio-temporal distribution has been linked to its preferred prey of squid, particularly *Todarodes sagittatus*, *Gonatus* and *Illex* spp., although fish such as Atlantic mackerel and shrimps may also be taken seasonally (Mercer 1975; Evans 1980; Waring *et al*. 1990).

HARBOUR PORPOISE

Phocoena phocoena

The harbour or common porpoise is the smallest and by far the most numerous of the cetaceans found in north-western European continental shelf waters. It has a small, rotund body with a short, blunt head, no beak, and a small, triangular dorsal fin. The species varies in size with geographical location, and males are slightly smaller than females of the same age. In the North Sea, females grow to about 160 cm in length, males to around 145 cm; newborn calves are typically 70-80 cm in length (Van Utrecht 1978; Lockyer 1995). Porpoises in Iberian waters and those south to Senegal are generally larger (Fraser 1958,; Smeenk et al. *1992; Sequeira 1996; IWC 1996).*

Typically, porpoises occur in small groups of one to three animals. Under certain circumstances they may form large aggregations but these do not appear to be co-ordinated schools, and probably result from many small groups and individuals concentrating at the same place at the same time for the same reason, for example to exploit good feeding resources (Hoek 1992).

The diet of the harbour porpoise comprises small fish of a wide variety of species (Read 1999). In the north-east Atlantic, small gadoids such as whiting, poor cod and Norway pout predominate, while herring, sandeels and gobies may be important at certain times or locations (Rae 1973; Santos Vázquez 1998).

Global distribution

The harbour porpoise occurs primarily in temperate waters of the North Pacific and North Atlantic, mainly but not exclusively over the continental shelves of these regions. In the eastern Atlantic, its distribution ranges from the Russian White Sea, in the summer at least, through most of the European shelf south to Senegal, with the most southerly records from just south of Cap Vert in Senegal (15° S). In the western Atlantic, porpoises have been recorded as far north as Upernavik in West Greenland (72° N), and as far south as Florida in the USA. In the Pacific, the species is found from northern Japan through the Aleutian Islands chain to northern Alaska, seasonally as far east as the MacKenzie River delta, and also along the entire North American shelf from Alaska as far south as Monterey Bay (IWC 1996).

North Atlantic status

The harbour porpoise is the most numerous marine mammal in north-western European shelf waters. There have been several estimates of population numbers in different parts of this area, but the SCANS survey in July 1994 (Hammond *et al.* 1995) has been the most wide-ranging. The North Sea population was estimated at about 280,000 animals, with a further 36,000 in the Skagerrak and Belt Seas and another 36,000 over the Celtic shelf between Ireland and Brittany.

Surveys around Norway have resulted in estimates of about 11,000 porpoises in waters north of 66° N and the Barents Sea, and 82,000 for the northern North Sea and southern Norwegian waters (Bjørge and Øien 1995). In inner Danish waters, Heide-Jørgensen *et al.* (1993) estimated around 500-580 porpoises to the North of Fyn, just over 500 in the Great Belt and just under 100 in the Little Belt in June 1991 and June 1992. The same authors estimated around 90-200 in these two years for the Kiel Bight, and somewhere between 100 and 500 around the Island of Sylt. Elsewhere, Leopold *et al.* (1992) estimated that there were around 19,000 harbour porpoises on the continental shelf off south-west Ireland.

In certain areas, populations seem to have declined or have been eliminated, notably in the eastern Channel and southern North Sea (Addink and Smeenk 1999), the Black and the Baltic Seas. Porpoises were clearly abundant in the Baltic until some time in the 1960s, when they appear to have declined (Koschinski 2002).

It is thought that levels of accidental mortality in fishing nets in certain areas might be unsustainable (ASCOBANS 2000).

Tooth ultrastructure and genetic studies indicate population differentiation in the North Sea and adjacent waters, with possible sub-populations around the British Isles in the Irish Sea and Wales, in the northern North Sea, eastern (Denmark) and western (British Isles) North Sea and southern North Sea (the Netherlands) (Tiedemann 1996; Walton 1997; Evans *et al.* in press).

NW European distribution

In the north-western European region, the harbour porpoise is mainly confined to shelf waters, although sightings have occurred in deep water, for example between the Faroe Islands and Iceland, suggesting some movement between adjacent shelf areas. Areas of highest population density appear to be in the Belt Sea to the east of Denmark and in the north-western North Sea, in water shallower than about 100 m. The southern, and especially the south-eastern, North Sea hosts relatively lower densities on an annual basis, and there have been few sightings in the Channel. On the Atlantic seaboard, there appear to be locally high densities of porpoises, such as off south-west Ireland and south-west Wales, and off the west coast of Scotland.

Seasonal movements are difficult to infer from the rather patchy monthly survey effort achieved. The highest sightings rates in the south-eastern North Sea, where there are relatively few sightings for the year as a whole, are during the first four months of the year, whereas the highest rates around the Outer Hebrides appear to be during summer (June to September). Whether such observations imply anything about seasonal movements, or are simply the results of differences in sightings efficiency among the various surveys involved, is not clear.

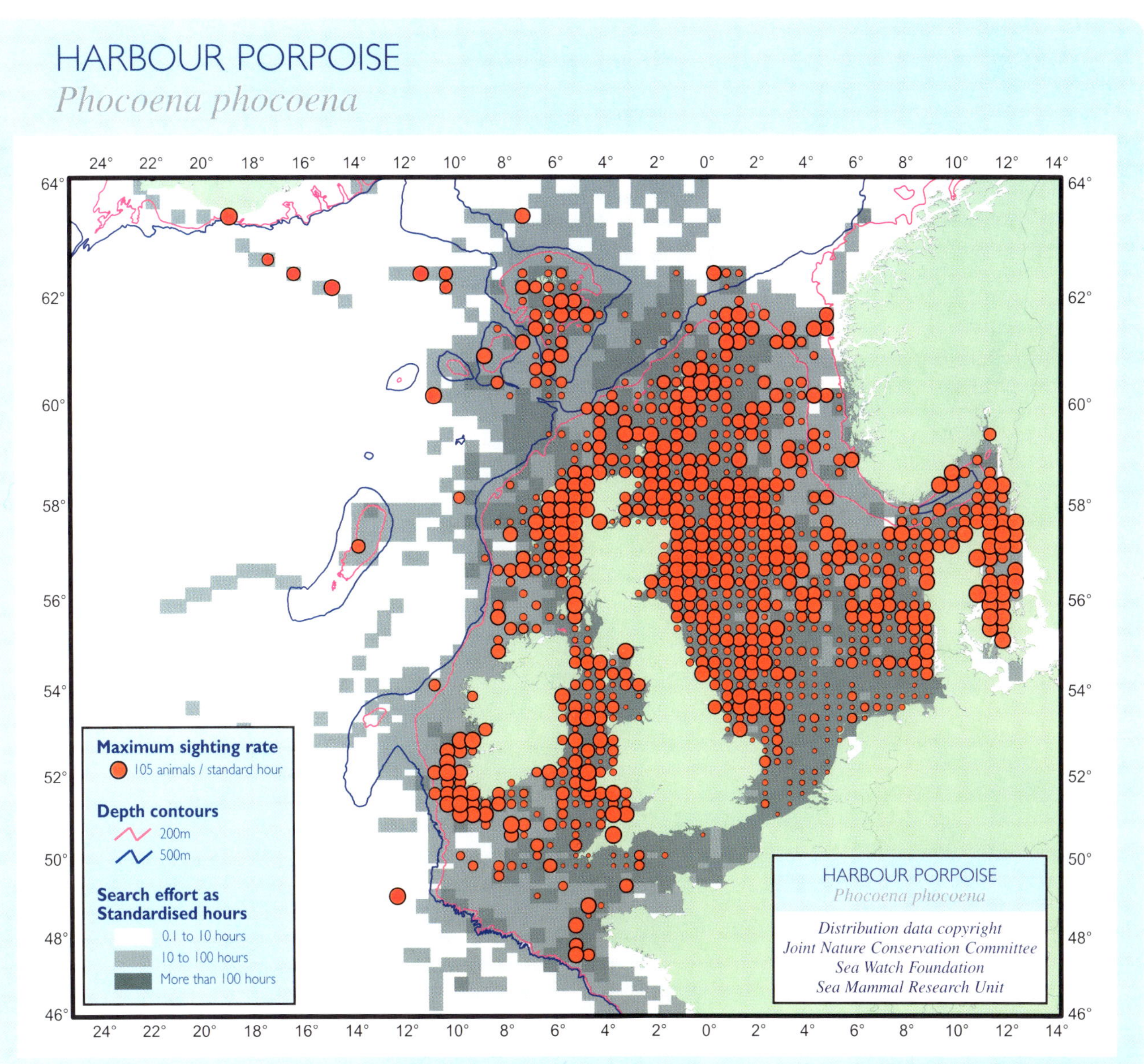

INTERNATIONAL PROTECTION OF CETACEANS

A number of international treaties, agreements and legislative instruments have been established, particularly since 1990, which either provide a legal basis for the conservation and protection of marine wildlife, including cetaceans, or represent statements of intent through agreements. Some of these are regional in their scope while others have global application. The main ones that pertain to waters covered by the Atlas are listed below.

Regional Conventions, Agreements and Legislation

Bern Convention

(1979; implemented 1982). The Convention on the Conservation of European Wildlife and Natural Habitats is concerned with the protection of natural habitats in Europe, especially the endangered habitats and wild flora and fauna that are listed in two Appendices. Appendix II accords strict protection to common dolphin, bottlenose dolphin, harbour porpoise, blue whale, humpback whale, northern right whale and bowhead whale. Appendix III lists species of wild fauna that may be exploited so long as regulation keeps the populations out of danger, and includes all other cetaceans. The Emerald Network (1996) of Areas of Special Conservation Interest (ASCI; 1989), which applies to all European countries including those bordering the Baltic and Black Seas, was established under the Bern Convention.

Bonn Convention

(1979; implemented 1983). The Convention on the Conservation of Migratory Species (CMS) provides strict protection for 28 endangered migratory species listed in Appendix I. Among cetaceans, this includes blue whale, humpback whale, bowhead whale and northern right whale. It also concludes multilateral agreements for the conservation and management of migratory species listed in Appendix II (i.e. those species considered to have unfavourable conservation status, including all other cetaceans), and promotes the undertaking of co-operative research activities. ASCOBANS (see below) was negotiated under its auspices.

EU Habitats and Species Directive (1992)

Council Directive 92/43/EEC on the conservation of natural habitats and of wild flora and fauna applies to marine waters out to the 200 nautical mile limits of the Economic Exclusion Zones (EEZ) of European Union member states. Member states are legally obliged to observe the Directive. Two species, harbour porpoise and bottlenose dolphin, are listed in Annex 2, which comprises 'animal and plant species of Community interest whose conservation require the designation of Special Areas of Conservation'. Sites for these two wide-ranging aquatic species will be proposed only where 'there is a clearly identifiable area representing the physical and biological factors essential to their life and reproduction'. There is an obligation on member states to avoid deterioration at such sites. Annex 4 includes all cetaceans in a list of animal and plant species of Community interest in need of strict protection. Member states of the European Union are required to establish a system to monitor the incidental capture and killing of all cetaceans and to take measures to ensure that such capture and killing does not have a significant negative impact on the species concerned.

Other EU Directives and Regulations

A number of other EU Directives, regulations and policies are of direct relevance to the protection of cetaceans. Among them are the Regulations controlling fishing under the Common Fisheries Policy. Regulation 2371/2002 requires that fishing should not have a negative impact on marine ecosystems and various other measures are required when taking decisions on fisheries management to reduce impact on the marine environment.

OSPAR

(Oslo and Paris Convention 1992; implemented 1998). The Convention for the Protection of the Marine Environment of the Northeast Atlantic provides in Annex V for the 'protection and conservation of the ecosystem and biological diversity of the maritime area', and in Appendix 3 lays down 'criteria for identifying human activities for the purpose of Annex V'. Seven species of cetacean are listed in Annex V. Four species are proposed for inclusion on OSPAR's first list of threatened and declining species: bowhead whale, blue whale, northern right whale and harbour porpoise.

ASCOBANS

(1992; implemented 1994). The Agreement on the Conservation of Small Cetaceans of the Baltic and North Seas was established under the Bonn Convention CMS (see above). It applies to all odontocetes except the sperm whale within the area of the Agreement, and obliges signatories to apply the conservation, research and management measures prescribed in its Annex that provide for monitoring, research, information exchange, bycatch reduction, pollution control, and heightening public awareness.

Global Conventions and Agreements

International Convention for the Regulation of Whaling

(1946; implemented 1948, with a protocol of amendment to the Convention adopted in 1956). This convention aims inter alia to 'provide for the proper conservation of whale stocks and thus make possible the orderly development of the whaling industry'. It established the International Whaling Commission (IWC). Although concerned primarily with the direct exploitation and conservation of large whales (the baleen whales and the sperm whale) as a resource, it also considers scientific matters related to small cetaceans, and has more recently begun to address other human activities affecting cetaceans. A moratorium on commercial whaling came into force in 1986. The only commercial whaling currently undertaken is by Norway, which objected to the moratorium within a prescribed period and so is not bound by it. Other whaling that is regulated by the IWC comes under the heading of aboriginal subsistence whaling. However, some hunting is not regulated, including Faroese pilot whaling.

Convention on International Trade in Endangered Species of Wild Fauna and Flora - CITES

(1973; implemented 1975). This prohibits international trade in species listed in Appendix 1, which includes sperm whale, northern right whale, and members of the family Balaenopteridae (including blue, fin, sei, and humpback whales, but with the exception of West Greenland populations of humpback and minke whales). It allows controlled trade in all other cetacean species.

MARPOL Agreement

(1973/78; implemented 1983). The 1973 International Convention on the Prevention of Pollution from Ships and 1978 Protocol Relating Thereto covers all technical aspects of pollution from ships of all types: accidental and operational oil pollution, pollution by chemicals, goods in packaged form, sewage and garbage. Parties are obliged to give effect to the provisions of the Convention and its Annexes in order to prevent the pollution of the marine environment by the discharge of harmful substances or effluents containing such substances in contravention of the Convention.

UNCLOS

(1982; implemented 1995). The UN Convention on the Law of the Sea, among other requirements, establishes fundamental obligations including the preservation and protection of the marine environment and the conservation of marine living resources both within and beyond national jurisdiction. Under UNCLOS, coastal states can extend their jurisdiction to their Exclusive Economic Zone (EEZ).

Earth Summit

(1992; implemented 1994). The Convention on Biological Diversity (CBD) requires each signatory to identify processes and categories of activities that are likely to have significant adverse impacts on the conservation and sustainable use of biological diversity, and to monitor their effects through sampling and other techniques. This should involve the introduction of appropriate procedures requiring environmental impact assessment of proposed potentially detrimental projects with a view to avoiding or minimising such effects. Parties are obliged to adopt in-situ measures to rehabilitate and restore degraded ecosystems and to promote the recovery of threatened species; to endeavour to provide the conditions needed for compatibility between present uses and the conservation of biological diversity and the sustainable use of its components; and to regulate or manage the relevant processes and categories of activities where a significant effect on biological diversity has been determined.

ACKNOWLEDGEMENTS

Thousands of observers contributed data to the separate databases that comprise the Joint Cetacean Database.

Many thanks to them.

We thank the various organisations who contributed data to the ESAS database: Nederlands Instituut voor Onderzoek der Zee (the Netherlands), Ornis Consult (Denmark), National Institute for Coastal and Marine Management/RIKZ (the Netherlands), Nederlands Zeevogelgroep (the Netherlands), Instituut voor Bos- en Natuuronderzoek (the Netherlands), Instituut voor Natuur Behoud (Belgium), National Environmental Research Institute (Denmark), Norsk Institutt for Naturforskning (Norway), Vogelwarte Helgoland (Germany). In particular, we are grateful to Kees Camphuysen, Mardik Leopold, Stefan Garthe, Ommo Hüppop and Henk Offringa.

Jim Boran and Caroline Weir made very important contributions to the Sea Watch database, both in terms of management and data input. Of the very many observers and organisations who contributed effort-related sightings data towards the Sea Watch database, the following deserve special mention: Biscay Dolphin Research Programme (Tom Brereton, Andy Williams and Rolf Williams), Cornwall Wildlife Trust (Ray Dennis, Lance Peters and Nick Tregenza), Earthkind (Gillian Bell, Christina Thomas and Peter Todd), Friends of Cardigan Bay (Mick Green and Megan Morgan-Jenks), Plymouth-Santander ferry surveys (Neil Fletcher and Dave Curtis), West Wales Chartering Company (Steve Hartley and observers), Western Isles Sailing (Christopher Swann and Teo Leyssen), and all the Sea Watch regional groups,

co-ordinators and observers who have made effort-related observations (especially Mick Baines, Dave Ball, David Barker, Cliff Benson, Simon Berrow, David Bird, Ian Birks, Chris Booth, Bertram Bree, Terry Bristow, Brian Burnett, Sarah Canning, Peter Cheeseman, Rob Colley, Ian Cumming, John Cudworth, Peter Dare, Linda Denton, Richard Dixon, Tony Douglas, Sarah Earll, Tony Elvers, Richard and Brennan Fairbairns, Shay Fennelly, Paul Fisher, Jim Fitzharris, Judy Foster-Smith, David Galloway, Martin Gavet, Lucy Gilbert, John Hagan, Norman Hammond, Rachel Harding-Hill, Hugh Harrop, John Hartley, Paul Harvey, Hebridean Whale & Dolphin Trust, Kevin Hepworth, Martin Heubeck, Iain Hill, Ian Hotchin, Mike Innes, David Jenkins, Ian Kendall, Richard Lansdown, Pete Leeming, Emily Lewis, Howard Loates, Iain Andrew Macdonald, Peter & Lyn Macdonald, Barbara Manson, Jo Le Marquand, Kenny Meek, Cathryn Owen, Robin Petch and Kris Simpson, Chris Pierpoint, Keith Pyman, Lesley Sinclair, Ken Preston, Martin Rosen, Steve Savage, Frank Scampton, Scottish Wildlife Trust wardens, Richard Shucksmith, Nigel Smallbones, Colin Speedie, Karen Stockin, Andy Tait, Dave Thomas, Nick Tregenza, Malcolm Turnbull, Jane Twelves, Andy Upton, David Walker, John Walton, Sue Warne, Ken Watterson, Caroline Weir, Padraig Whooley, Stuart Wright, Paul and Grace Yoxon, and numerous wardens of coastal bird observatories and reserves).

A great many organisations in the public, private, charity and education sectors have sponsored the collection of the data that are presented in the Atlas. We extend continued thanks to them, but in particular Sea Watch would like to acknowledge funding support from the Esmée Fairbairn Trust, BG plc and the UK Department of the Environment. The SCANS survey was funded by the European Commission LIFE programme under project LIFE 92-2/UK/027.

Chris Smeenk of the National Museum of Natural History in Leiden, the Netherlands, read most of the book and offered constructive criticism of it that we hope has resulted in improvements in the text. Vin Fleming, Wyn Jones and Mark Tasker of the JNCC read parts of the text and made useful suggestions and additions to it.

Mark Bravington (currently at CSIRO in Hobart) provided advice and the sea state correction factors used in the production of the Atlas. Andy Webb of the JNCC expertly maintains the ESAS database and ensured its use in this project was largely trouble-free. Daniel Exeter of the Geography Department at the University of St Andrews contributed invaluable assistance in producing the maps.

Thanks once again to all the contributors to the species chapters whose names appear on the title page.

Alex Geairns of the JNCC ensured publication of the Atlas and contributed greatly to its final form. Status Design and Advertising gave the vision form.

REFERENCES

Addink, M J and Smeenk, C (1999) The harbour porpoise *Phocoena phocoena* in Dutch coastal waters: analysis of stranding records for the period 1920-1994. *Lutra*, **41**, 55-80

Amos, B, Schlotterer, C and Tauz, D (1993) Social structure of pilot whales revealed by analytical DNA profiling. *Science*, **260**, 670-672

Arnold, H, Bartels, B, Wilson, B and Thompson, P (1997) The bottlenose dolphins of Cardigan Bay: Selected biogra phies of individual bottlenose dolphins recorded in Cardigan Bay, Wales. *Report to Countryside Council for Wales*. CCW Science Report No. 209

ASCOBANS (2000) *Agreement on the Conservation of Small Cetaceans of the Baltic and North Seas*. Proceedings of the Third meeting of Parties to ASCOBANS. Bristol, UK, 26-28 July 2000

ASCOBANS (2002). Recovery Plan for Baltic Harbour Porpoises (Jastarnia Plan). Bonn, ASCOBANS

Atkinson, T, Gill, A, and Evans, P G H (1999) A photo-identification study of Risso's dolphins in the Outer Hebrides, Northwest Scotland. *European Research on Cetaceans*, **12**, 102

Baird, R W and Stacey, P J (1991) Status of Risso's dolphin, *Grampus griseus*, in Canada. *Canad. Field-Natur.*, **105**, 233-242

Baird, R W, Abrams, P A and Dill, L M (1992) Possible indirect interactions between transient and resident killer whales: implications for the evolution of foraging specialization in the genus *Orcinus*. *Oecologia*, **89**, 125-132

Baird, R W, Langelier, K M and Stacey, P J (1989) First records of false killer whales *(Pseudorca crassidens)* in Canada. *Canad. Field-Natur.*, **103**, 368-371

Bakke, I, Johanse, S, Bakke, O and El-Gewely, R (1996) Lack of population subdivision among the minke whales from Icelandic and Norwegian waters based on mito chondrial DNA sequences. *Mar. Biol.*, **125**, 1-9

Balcomb, K C, Boran, J R and Heimlich, S L (1982) Killer whales in Greater Puget Sound. *Rep. Int. Whal. Commn*, **32**, 681-685

Barlow, J and Clapham, P J (1997) A new birth-interval approach to estimating demographic parameters of humpback whales. *Ecology*, **78**, 535-546

Benjaminsen. T and Christensen, I (1979) The natural history of the bottlenose whale, *Hyperoodon ampullatus* Forster. In *Behavior of Marine Mammals* (eds H E Winn and B L Olla), pp. 143-164. Plenum Press, New York

Bernard, H J and Reilly, S B (1999) Pilot whales *Globicephala* Lesson, 1828. In *Handbook of marine mammals. Volume 6. The second book of dolphins and the porpoises* (eds S H Ridgway and R Harrison), pp. 245-279. Academic Press, London

Berrow, S, Evans, P G H and Sheldrick, M (1993) An analysis of sperm whale *Physeter macrocephalus* stranding and sighting records from Britain and Ireland. *J. Zool.*, Lond., **230**, 333-337

Berrow, S D and Rogan, E (1995) Stomach contents of harbour porpoise and dolphins in Irish waters. *European Research on Cetaceans*, **9**, 179-181

Berrow, S D and Rogan, E (1997) Cetaceans stranded on the Irish coast. *Mamm. Rev.*, **27**, 51-75

Bérubé, M, Aguilar, A, Dendanto, D, Larsen, F, Notarbartolo di Sciara, G, Sears, R, Sigurjónsson, J, Urban-Ramirez, J and Palsbøll, P (1998) Population genetic structure of North Atlantic, Mediterranean Sea and Sea of Cortez fin whales, *Balaenoptera physalus* (Linnaeus, 1758): analysis of mitochondrial and nuclear loci. *Molecular Ecology*, **7**, 585-599

Bigg, M A, Olesiuk, P F, Ellis, G M, Ford, J K B and Balcomb, K C (1990) Social organization and genealogy of resident killer whales *(Orcinus orca)* in the coastal waters of British Columbia and Washington State. *Rep. Int. Whal. Commn* (special issue 12), 383-405

Bjørge, A and Øien, N (1995) Distribution and abundance of harbour porpoise, *Phocoena phocoena* in Norwegian waters. *Rep. Int. Whal. Commn* (special issue 16), 89-98

Bloch, D (1998) A review of marine mammals observed, caught or stranded over the last two centuries in Faroese waters. *Shetland Sea Mammal Report*, 1997, 15-37

Bloch, D, Desportes, G, Zachariassen, M and Christensen, I (1996) The northern bottlenose whale in the Faroe Islands. *J. Zool.*, Lond., **239**, 123-140

Bloor, P, Reid, J, Webb, A, Begg, G and Tasker, M (1996) *The distribution of seabirds and cetaceans between the Shetland and Faroe Islands*. JNCC Report, No. 226.

Joint Nature Conservation Committee, Peterborough

Bones, M, Kitchener, A and Reid, R J (1998) Fraser's dolphin *(Lagenodelphis hosei)* stranded in South Uist: first record in U.K. waters. *J. Zool.*, Lond., **246**, 460-461

Boran, J R, Evans, P G H and Rosen, M R (2002) Behavioural ecology of cetaceans. In *Marine Mammals: Biology and Conservation* (eds P G H Evans and J A Raga), pp. 197-242. Plenum Press/Kluwer Academic, London

Bravington, M V, Northridge, S P, Webb, A (2001) An exploratory analysis of cetaceans sightings data col lected from platforms of opportunity. Unpublished report to MAFF

Bravington, M, Borchers, D and Northridge, S (2002) Analysis of harbour porpoise sightings data in relation to area-based conservation. Unpublished report to JNCC

Brown, S G (1976) Modern whaling in Britain and the North-east Atlantic Ocean. *Mamm. Rev.*, **6**, 25-36

Brown, S G (1986) Twentieth century records of right whales *(Eubalaena glacialis)* in the northeast Atlantic Ocean. *Rep. Int. Whal. Commn* (special issue 10), 121-127

Brownell, R L, Jr and Ralls, K (1986) Potential for sperm competition in baleen whales. *Rep. Int. Whal. Commn* (special issue 8), 97-112

Bryden, M M, Harrison, R J and Lear, R J (1977) Some aspects of the biology of *Peponocephala electra* (Cetacea: Delphinidae). I. General and reproductive biology. *Austr. J. Mar. Freshw. Res.*, **28**, 703-715

Buckland, S T, Anderson, D R, Burnham, K P, Laake, J L, Borchers, D L and Thomas, L (2001) *Introduction to Distance Sampling. Estimating abundance of biological populations.* Oxford University Press, Oxford

Buckland, S T, Bloch, D, Cattanach, K L, Gunnlaugsson, T, Hoydal, K, Lens, S and Sigurjónsson, J (1993) Distribution and abundance of long-finned pilot whales in the North Atlantic, estimated from NASS-87 and NASS-89 data. Rep. Int. Whal. Commn (special issue 14), 33-49

Buckland, S T, Cattanach, K L and Gunnlaugsson, Th. (1992) Fin whale abundance in the eastern North Atlantic, estimated from Spanish NASS-89 data. *Rep. Int. Whal. Commn*, **42**, 457-460

Budylenko, G A (1977) Distribution and composition of sei whale schools in the southern hemisphere. *Rep. Int. Whal. Commn* (special issue 1), 121-123

Cagnolaro, L, Di Natale, A and Notarbartolo di Sciara, G (1983) *Cetacei. Guide per il riconoscimento delle specie animali delle acque lagunari e costiere italiane.* 9. CNR, Rome

Caldwell, D K and Caldwell, M C (1989) Pygmy sperm whale *Kogia breviceps* de Blainville, 1838: Dwarf sperm whale *Kogia simus* Owen, 1866. In *Handbook of Marine Mammals. Volume 4. River dolphins and the larger toothed whales* (eds S H Ridgway and R Harrison), pp. 235-260. Academic Press, San Diego, CA

Cañadas, A M and Sagarminaga, R (1996) Preliminary results of photo-identification studies on Risso's dolphins *(Grampus griseus)* undertaken during surveys of cetacean distribution and dynamics along the south-east coast of Spain: 1992-1995. *European Research on Cetaceans*, **10**, 221-224

Carlisle, A, Coles, P, Cresswell, G, Diamond, J, Gorman, L, Kinley, C, Rondel, G, Sellwood, D, Still, C, Telfer, M and Walker, D (2001) A report on the whales, dolphins and porpoises of the Bay of Biscay and English Channel 1999. *Orca*, **1**, 5-53

Carlstrom, J, Denkinger, J, Fedderson, P and Øien, N (1997) Record of a new northeren range of Sowerby's beaked whale *(Mesoplodon bidens)*. *Polar Biology*, **17**, 459-461

Carrillo, M and Lopez-Jurado, L F (1998) Structure and behaviour of a Blainville's beaked whale *(Mesoplodon densirostris)* group in Tenerife (Canary Islands). *European Research on Cetaceans*, **12**, 74

Carrillo, M and Martín, V (1999) First sighting of Gervais' beaked whale (*Mesoplodon europaeus* Gervais, 1855) (Cetacea; Ziphiidae) from the North Oriental Atlantic coast. *European Research on Cetaceans*, **13,** 180

Castells, A and Mayo, M (1992) The Cuvier's beaked whale, *Ziphius cavirostris*, in the Iberian Peninsula. *European Research on Cetaceans*, **6**, 94-96

Cattanach, K L, Sigurjónsson, J, Buckland, S T and Gunnlaugsson, Th. (1993) Sei whale abundance, esti mated from Icelandic and Faroese NASS-87 and NASS-89 data. *Rep. Int. Whal. Commn*, **43**, 315-321

Charif, R A and Clark, C W (2000) Acoustic monitoring of large whales off north and west Britain and Ireland: a two-year study, October 1996-September 1998. JNCC Report No. 313

Christensen, I, Haug, T and Wiig, Ø. (1990) Morphometric comparison of minke whales *(Balaenoptera acutorostrata)* from different areas of the North Atlantic. *Mar. Mamm. Sci.*, **6**, 327-338.

Clapham, P J (1996) The social and reproductive biology of Humpback Whales: an ecological perspective. *Mamm. Rev.*, **26**, 27-50

Clarke, M R (1986) Cephalopods in the diet of odonto cetes. In *Research on Dolphins* (eds M M Bryden and

R Harrison), pp. 281-322. Clarendon Press, Oxford

Clarke, M R and Kristensen, T K (1980) Cephalopod beaks from the stomachs of two northern bottlenosed whales *(Hyperoodon ampullatus)*. *J. Mar. Biol. Assoc. U.K.*, **60**, 151-156

Clarke, M R and Pascoe, P L (1985) The stomach contents of a Risso's dolphin *(Grampus griseus)* stranded at Thurlestone, South Devon. *J. Mar. Biol. Assoc. U.K.*, **65**, 663-665

Coles, P, Diamond, J, Harrop, H, MacLeod, K and Mitchell, J. (2001) A report on the whales, dolphins and porpoises of the Bay of Biscay and English Channel 2000. *Orca*, **2**, 9-61

Collet, A (1981) *Biologie du Dauphin commun Delphinus delphis L. en Atlantique Nord-Est*. PhD Thesis, University of Poitiers

Collet, A and Evans, P G H (in press) The striped dolphin *Stenella coeruleoalba*. In *New Handbook of British Mammals* (ed. S Harris). Academic Press, London

Collet, A, Wilson, B and Evans, P G H (in press) Fraser's dolphin *Lagenodelphis hosei*. In *New Handbook of British Mammals* (ed. S Harris). Academic Press, London

Collett, R (1909) A few notes on the whale *Balaena glacialis* and its capture in recent years in the North Atlantic by Norwegian whalers. *Proc. Zool. Soc. Lond.*, **7**, 91-97

Connor, R C, Smolker, R and Richards, A F (1992) Two levels of alliance formation among male bottlenose dolphins (*Tursiops* spp.). *Proc. Natl. Acad. Sci. USA*, **89**, 987-990

Couperus A S (1997) Interactions between Dutch midwater trawl and Atlantic white-sided dolphins *(Lagenorhynchus acutus)* southwest of Ireland. *J. Northw. Atl. Fish. Sci.*, **2**, 209-218

Couperus, A S (1999) Diet of Atlantic white-sided dolphins southwest of Ireland. *European Research on Cetaceans*, **12**, 107

Couperus, B (1993) Killer whales and pilot whales near trawlers east of Shetland. *Sula*, **7**, 41-52

Dahlheim, M E and Heyning, J E (1999) Killer Whale, *Orcinus orca*. In *Handbook of Marine Mammals. Vol. 6. The second book of dolphins and the porpoises* (eds S H Ridgway and R Harrison), pp. 281-322. Academic Press, London

Daníelsdóttir, A K, Halldórsson, S D, Gulaugsdóttir, S and Árnason, A (1995) Genetic variation in northeastern Atlantic minke whales *(Balaenoptera acutorostrata)*. In *Whales, seals, fish and man* (eds A S Blix, L Walløe and Ø. Ulltang), pp. 105-118. Elsevier Press, Amsterdam

Desportes, G (1985) *La nutrition des Odontocètes enAtlantique Nord-Est (côtes françaises & îles Féroë)*. PhD Thesis, University of Poitiers

Desportes, G and Mouritsen, R (1993) Preliminary results on the diet of long-finned pilot whales off the Faroe Islands. *Rep. Int. Whal. Commn* (special issue 14), 305-324

Desportes, G, Saboureau, M and Lacroix, A (1993) Reproductive maturity and seasonality of male long-finned pilot whales, off the Faroe Islands. *Rep. Int. Whal. Commn* (special issue 14), 233-262

Dietz, R and Heide-Jørgensen, M P (1995) Movements and swimming speed of narwhals, *Monodon monoceros*, equipped with satellite transmitters in Melville Bay, N.W. Greenland. *Can. J. Zool.*, **73**, 2106-2119

Dorsey, E M, Stern, S J, Hoelzel, A R and Jacobsen, J (1990) Minke whales *(Balaenoptera acutorostrata)* from the west coast of North America: individual recognition and small-scale site fidelity. *Rep. Int. Whal. Commn* (special issue 12), 357-368

Duguy, R (1966) Quelques données nouvelles sur un Cétacé rare sur les côtes d'Europe: Le Cachalot à tête courte, *Kogia breviceps* (Blainville, 1838). *Mammalia*, **30**, 259-269

Edds, P L and Macfarlane, J A F (1987) Occurrence and gen eral behavior of balaenopterid cetaceans summering in the St. Lawrence Estuary, Canada. *Can. J. Zool.*, 65, 1363-1376

Ellett, D J, Krauseman, P, Prangsma, G J, Pollard, R T, Van Aken, H M, Edwards, A, Dooley, H D and Gould, W J 1983. Water masses and mesoscale circulation of North Rockall Trough waters during JASIN 1978. *Philosophical Transactions of the Royal Society of London, A*, **308**, 231-252

Evans, P G H (1980) Cetaceans in British waters. *Mamm. Rev.*, **10**, 1-52

Evans, P G H (1987) *The Natural History of Whales and Dolphins*. Christopher Helm/Academic Press, London

Evans, P G H (1988) Killer whales *(Orcinus orca)* in British and Irish waters. *Rit Fiskid.*, **11**, 42-54

Evans, P G H (1991) Whales, Dolphins and Porpoises. Order Cetacea. In *The Handbook of British Mammals* (eds G B Corbet and S Harris), pp. 299-350. Blackwell, Oxford

Evans, P G H (1992) *Status Review of Cetaceans in British and Irish Waters*. Report to UK Dept. of the Environment. UK Mammal Society Cetacean Group, Oxford

Evans, P G H (1996) Humpback Whales in Shetland. *Shetland Cetacean Report*, 1995, 7-8

Evans, P G H (1996) Sightings frequency and distribution of

cetaceans in Shetland waters. *The Shetland Cetacean Group Report*, 1995, 9-18

Evans, P G H (1997) Ecology of Sperm Whales *(Physeter macrocephalus)* in the Eastern North Atlantic, with special reference to sightings and strandings records from the British Isles. In *Sperm Whale Deaths in the North Sea: Science and Management* (eds T G Jacques and R H Lambertsen), pp. 37-46. *Bull. de L'Institut Royal des Sciences Naturelles de Belgique. Biologie. Vol. 67 - Supplement*

Evans, P G H (in press) Risso's Dolphin *Grampus griseus*. In *New Handbook of British Mammals* (ed. S Harris). Academic Press, London

Evans, P G H and Smeenk, C (in press) Atlantic White-sided Dolphin *Lagenorhynchus acutus*. In *New Handbook of British Mammals* (ed. S Harris). Academic Press, London

Evans, P G H and Smeenk, C (in press) White-beaked Dolphin *Lagenorhynchus albirostris*. In *New Handbook of British Mammals* (ed. S Harris). Academic Press, London

Evans, P G H and Wang, J (2003) *Re-examination of distribution data for the harbour porpoise around Wales and the UK with a view to site selection for this species*. Unpublished Report to Countryside Council for Wales

Evans, P G H, Lockyer, C and Read, A (in press) Harbour porpoise *Phocoena phocoena*. In *New Handbook of British Mammals* (ed. S Harris). Academic Press, London

Evans, W E (1994) Common dolphin, White-bellied porpoise *Delphinus delphis* Linnaeus, 1758. In *Handbook of Marine Mammals. Vol. 5. The first book of dolphins* (eds S H Ridgway and R Harrison), pp. 191-224. Academic Press, London

Fabbri, F, Giordano, A and Lauriano, G (1992) A preliminary investigation into the relationship between the distribution of Risso's dolphin and depth. *European Research on Cetaceans*, **6**, 146-151

Fairley, J S (1991) *Irish Whales and Whaling*. Blackstaff Press, Belfast

Faucher, A and Whitehead, H (1991) *The bottlenose whales of the Gully*. Final Report for 1988-91 Project to WWF Canada. Unpublished report. World Wildlife Fund - Canada, Toronto

Felleman, F L, Heimlich-Boran, J R and Osborne, R W (1991) Feeding ecology of the killer whale *(Orcinus orca)*. In *Dolphin Societies* (eds K Pryor and K S Norris), pp. 113-147. University of California Press, Berkeley

Finley, K J and Gibb, E J (1982) Summer diet of the narwhal *(Monodon monoceros)* in Pond Inlet, northern Baffin Island. *Can. J. Zool.*, **60**, 3353-3363

Forcada, J, Aguilar, A, Evans, P G H and Perrin, W F (1990) Distribution of common and striped dolphins in the temperate waters of the eastern North Atlantic. *European Research on Cetaceans*, **4**, 64-66

Fraser, F C (1934) *Report on Cetacea stranded on the British coasts from 1927 to 1932*. British Museum (Natural History), London

Fraser, F C (1946) *Report on Cetacea stranded on the British coasts from 1933 to 1937*. British Museum (Natural History), London

Fraser, F C (1956) A new Sarawak dolphin. *Sarawak Mus. J.*, **7**, 478-503, pl. 22-26

Fraser, F C (1958) Common or harbour porpoise from French West Africa. *Bull. Inst. Fr. D'Afrique Noire, Serie A: Sci. Nat.*, **20**, 276-285

Fraser, F C (1974) *Report on Cetacea stranded on the British coasts from 1948 to 1966*. No. 14. British Museum (Natural History), London

Gambell, R (1979) The blue whale. *Biologist*, **26**, 209-215

Gambell, R. (1985) Sei Whale. In *Handbook of Marine Mammals. Volume 3: The Sirenians and Baleen Whales* (eds S H Ridgway and R Harrison), pp. 155-170. Academic Press, London

Gannier, A and David, L (1997) Day and night distribution of the striped dolphin *(Stenella coeruleoalba)* in the area off Antibes (Ligurian Sea). *European Research on Cetaceans*, **11**, 160-163

Gannier, A and Gannier, O (1994) Abundance of *Grampus griseus* in northwestern Mediterranean. *European Research on Cetaceans*, **8**, 99-102

Gill, A and Fairbairns, R S (1995) Photo-identification of the minke whale *Balaenoptera acutorostrata* off the Isle of Mull, Scotland. In *Whales, Seals, Fish and Man* (eds A S Blix, L Walløe and Ø Ulltang), pp. 129-132. Elsevier, Amsterdam

Gordon, J C D (1998) *Sperm whales*. Colin Baxter Photography, Grantown-on-Spey

Goujon, M (1996) *Captures accidentelles du filet maillant dérivant et dynamique des populations de dauphins au large du Golfe de Gascogne*. PhD Thesis, École nationale Supérieure Agronomique de Rennes

Goujon, M, Antoine, L, Collet, A and Fifas, S (1994) A study of the ecological impact of the French tuna driftnet fishery in the North-east Atlantic. *European Research on Cetaceans*, **8**, 47-48

Gunnlaugsson, T and Sigurjónsson, J (1990) NASS-87: estimation of whale abundance based on observations made on board Icelandic and Faeroese survey vessels. *Rep. Int. Whal. Commn*, **40**, 571-580

Hain, J H W, Edel, R K, Hays, H E, Katona, S K and Roanowicz, J D (1981) General distribution of cetaceans in the continental shelf waters of the N.E. United States. In Cetacean and Turtle Assessment Program. *A characterisation of marine mammals and turtles in the mid and North Atlantic areas of the US outer continental shelf*, pp. 11.1-11.345. Annual Report for 1979, contract AA551-CT8-48. US Dept. Int., Washington DC

Hammond, P S, Benke, H, Berggren, P, Borchers, D L, Buckland, S T, Collet, A, Heide-Jørgensen, M P, Heimlich-Boran, S, Hiby, A R, Leopold, M F and Øien, N (1995) *Distribution and abundance of the harbour porpoise and other small cetaceans in the North Sea and adjacent waters*. Final Report to the European Commission under contract LIFE 92-2/UK/27

Hammond, P S, Berggren, P, Benke, H, Borchers, D L, Buckland, S T, Collet, A, Heide-Jørgensen, M P, Heimlich, S, Hiby, A R, Leopold, M F and Øien, N (2002) Abundance of harbour porpoise and other cetaceans in the North Sea and adjacent waters. *J. Appl. Ecol.* **39**, 361-376

Hamner, W M and Haury, L R (1977) Fine-scale surface currents in the Whitsunday Islands, Queensland, Australia: effect of tide and topography. *Australian Journal of Marine and Freshwater Research*, **28**, 333-359

Harmer, S F (1924) On *Mesoplodon* and other beaked whales. *Proc. Zool. Soc. Lond.*, **94**, 541-587

Harwood, J, Andersen, L W, Berggren, P, Carlström, J, Kinze, C C, McGlade, J, Metuzals, K, Larsen, F, Lockyer, C H, Northridge, S, Rogan, E, Walton, M and Vinther, M (1999) Assessment and reduction of the by-catch of small cetaceans (BY-CARE). Final Report to the European Commission under contract FAIR-CT05-0523

Haug, T, Gjøsaeter, H, Lindstrøm, U, Nilssen, K T and Røtiingen, I (1995) Spatial and temporal variations in northeast Atlantic minke whale *Balaenoptera acutorostrata* feeding habits. In *Whales, Seals, Fish and Man* (eds A S Blix, L Walløe and Ø Ulltang), pp. 225-239. Elsevier, Amsterdam

Heide-Jørgensen, M P, Dietz, R and Leatherwood, S (1994) A note on the diet of narwhals *(Monodon monoceros)* in Inglefield Bredning (N.W. Greenland). *Meddelelserom Grønland, Bioscience,* **39**, 213-216

Heide-Jørgensen, M P and Dietz, R (1995) Some chararacteristics of narwhal, *Monodon monoceros*, diving behaviour in Baffin Bay. *Can. J. Zool.*, **73**, 2120-2132

Heide-Jørgensen, M P and Teilmann, J (1994) Growth, reproduction, age structure and feeding habits of white whales *(Delphinapterus leucas)* in West Greenland. *Meddelelser om Grønland, Bioscience*, **39**, 195-212

Heide-Jørgensen, M P, Teilmann, J, Benke, H and Wulf, J (1993) Abundance and distribution of harbour porpoises *(Phocoena phocoena)* in selected areas of the western Baltic and the North Sea. *Helgol. Meeresunters.*, **47**, 335-346

Heimlich-Boran, J R (1988) Behavioral ecology of killer whales *(Orcinus orca)* in the Pacific Northwest. *Can. J. Zool.*, **66**, 565-578

Herman, J S, Kitchener, A C, Baker, J R and Lockyer, C (1994) The most northerly record of Blainville's beaked whale, *Mesoplodon densirostris*, from the eastern Atlantic. *Mammalia*, **58**, 657-661

Heyning, J E (1989) Cuvier's beaked whale - *Ziphius cavirostris* G. Cuvier, 1823. In *Handbook of Marine Mammals. Vol. 4. River dolphins and the larger toothed whales* (eds S H Ridgway and R Harrison), pp. 289-308. Academic Press, London

Hill, A E (1993) Seasonal gyres in shelf seas. *Annales Geophysicae*, **11**, 1130-1137

Hill, A E and Simpson, J H 1989. On the interaction of thermal and haline fronts: the Islay front revisited. *Estuarine, Coastal and Shelf Science*, **28**, 495-505

Hoek, W (1992) An unusual aggregation of harbor porpoises *(Phocoena phocoena)*. *Mar. Mamm. Sci.*, **8**, 152-155

Hoelzel, A R (1991) Analysis of regional mitochondrial DNA variation in the killer whale; implications for cetacean conservation. *Rep. Int. Whal. Commn* (special issue 13), 224-233

Holligan, P M, Aarup, T and Groom, S B (1989) The North Sea Satellite Colour Atlas. *Continental Shelf Research*, **9**, 665-765

Hooker, S K and Baird, R W (1999) Deep-diving behaviour of the northern bottlenose whale, *Hyperoodon ampullatus* (Cetacea: Ziphiidae). *Proc. R. Soc. Lond.* B, **266**, 671-676

Hooker, S K and Baird, R W (1999) Observations of Sowerby's beaked whales, *Mesoplodon bidens*, in the Gully, Nova Scotia. Canad. *Field-Natur.*, **113**, 273-277

Horwood, J (1987) *The Sei Whale*. Academic Press, London

Houston, J (1991) Status of Cuvier's beaked whale, *Ziphius cavirostris*, in Canada. *Canad. Field-Natur.*, **105**(2), 215-218

Hoyt, E (1983) Great winged whales: combat and

courtship rites among humpbacks, the oceans' not so gentle giants. *Equinox*, **10**, 25-47

Hoyt, E (1990) *Orca: The whale called killer*. Robert Hayle Ltd, London

Huang, W G, Cracknell, A P, Vaughan, R A and Davies, P A (1991) A satellite and field view of the Irish Shelf Front. *Continental Shelf Research*, **11**, 543-562

Ingram, S, Rogan, E and O'Flanagan, C (1999) Population abundance and seasonal changes in abundance of resident bottlenose dolphins *(Tursiops truncatus)* in the Shannon Estuary, Ireland. *European Research on Cetaceans*, **13**, 328

Isaksen, K and Syvertsen, P O (2002) Striped dolphins, *Stenella coeruleoalba*, in Norwegian and adjacent waters. *Mammalia*, **66**, 33-41

IWC (1992) Report of the comprehensive assessment special meeting on north Atlantic fin whales. *Rep. Int. Whal. Commn*, **42**, 595-644

IWC (1996) Report of the Sub-Committee on Small Cetaceans. Annex H. *Rep. Int. Whal. Commn*, **46**, 160-179

Jacobsen, J (1986) The behavior of *Orcinus orca* in Johnstone Strait, British Columbia. In *Behavioural biology of killer whales* (eds B C Kirkevold and J S Lockard), pp. 135-185. A R Liss, New York

Jaquet, N (1996) How spatial and temporal scales influence understanding of sperm whale distribution: a review. *Mamm. Rev.*, **26**, 51-65

Jefferson, T A, Stacey, P J and Baird, R W (1991) A review of killer whale interactions with other marine mammals: predation to coexistence. *Mamm. Rev.*, **22**, 35-47

Jepson, P D and Baker, J R (1998) Bottlenosed dolphins *(Tursiops truncatus)* as a possible cause of acute traumatic injuries in porpoises *(Phocoena phocoena)*. *Vet. Rec.*, Nov. 28 1998, 614-615

Jonsgård, Å (1966) Biology of the North Atlantic fin whale *(Balaenoptera physalus)*. Taxonomy, distribution, migration and food. *Hvalraad Skr.*, **49**, 1-62

Jonsgård, Å and Darling, K. (1977) On the biology of the eastern North Atlantic sei whale, *Balaenoptera borealis* Lesson. *Rep. Int. Whal. Commn* (special issue 1), 121-123.

Jonsgård, Å and Øynes, P (1952) Om bottlenosen *(Hyperoodon rostratus)* og spekkhoggeren *(Orcinus orca)*. *Fauna*, Oslo, **5**, 1-18

Kasuya, T (1971) Consideration of distribution and migration of toothed whales off the Pacific coast of Japan based upon aerial sighting record. *Sci. Rept. Whales Res. I nst. Tokyo*, **23**, 37-60

Katona, S K and Kraus, S D (1999) Efforts to conserve the North Atlantic right whale. In *Conservation and Management of Marine Mammals* (eds J R Twiss and R R Reeves), pp. 311-331. Smithsonian Institution Press, Washington, DC

Kawakami, T (1980) A review of sperm whale food. *Sci. Rep. Whales Res. Inst. Tokyo*, **32**, 199-218

Kawamura, A (1980) A review of food of balaenopterid whales. Sci. Rep. *Whales Res. Inst., Tokyo*, **32**, 155-197

Kenney, R D, Hyman, M A M, Owen, R E, Scott, G P and Winn, H E (1986) Estimation of prey densities required by western North Atlantic right whales. *Mar. Mamm. Sci.*, **2**(1), 1-13

Kenney, R D and Winn, H E (1986) Cetacean high use habitats of the N.E. United States continental shelf. *Fish. Bull., U.S.*, **84**, 345-357

Kenney, R D and Winn, H E (1987) Cetacean biomass densities near submarine canyons compared to adjacent shelf/slope areas. *Contin. Shelf Res.*, **7**, 107-114

Kingsley, M C S, Cleator, M C S, Ramsay, H J and Ramsay, M A (1994) Summer distribution and movements of narwhals *(Monodon monoceros)* in Eclipse Sound and adjacent waters, North Baffin Island, NWT. *Meddr. om Grønland, Bioscience*, **39**, 163-174

Kinze, C C, Addink, M, Smeenk, C, García Hartmann, M, Richards, H W, Sonntag, R P and Benke, H (1997) The white-beaked dolphin *(Lagenorhynchus albirostris)* and the white-sided dolphin *(Lagenorhynchus acutus)* in the North and Baltic Seas: review of available information. *Rep. Int. Whal. Commn*, **47**, 675-681

Klinowska, M (1991) *Dolphins, Porpoises and Whales of the World*. The IUCN Red Data Book. IUCN, Gland, Switzerland

Knowlton, A R and Kraus, S D (2001) Mortality and serious injury of northern right whales *(Eubalaena glacialis)* in the western North Atlantic Ocean. *J. Cetacean Res. Manage*. Special Issue 2, 193-208

Koschinski, S. (2002) Current knowledge on harbour porpoises *Phocoena phocoena* in the Dutch sector of the North Sea. *Ophelia*, **55**, 167-197

Kraus, S D and Evans, P G H (in press). Northern right whale, *Eubalaena glacialis*. In *New Handbook of British Mammals* (ed. S Harris). Academic Press, London

Kraus, S D, Hamilton, P K, Kenney, R D, Knowlton, A R and Slay, C K (2001) Reproductive parameters of the North Atlantic Right Whale. *J. Cetacean Res. Manage*. Special Issue 2, 231-236

Kruse, S (1989) *Aspects of the biology, ecology and behaviour of Risso's dolphin (Grampus griseus) off the Californian coast*. MSc thesis, University of

California

Kruse, S, Caldwell, D K and Caldwell, M C (1999) Risso's dolphin *Grampus griseus* (G. Cuvier, 1812). In *Handbook of Marine Mammals. Vol. 6. The second book of dolphins and porpoises* (eds S H Ridgway and R Harrison), pp. 183-212. Academic Press, London

Leatherwood, J S and Reeves, R R (1983) *The Sierra Club Handbook of Whales and Dolphins*. Sierra Club Books, San Francisco

Leatherwood, S, Perrin, W F, Kirby, V L, Hubbs, C L and Dahlheim, M (1980) Distribution and movements of Risso's dolphin, *Grampus griseus*, in the eastern North Pacific. *Fish. Bull., U.S.*, **77**, 951-963

Leatherwood, S, Reeves, R R, Perrin, W F and Evans, W E (1988) *Whales, dolphins and porpoises of the eastern North Pacific and adjacent arctic waters: a guide to their identification*. Dover Publications, New York

Lee, A J and Ramster, J W 1981. *Atlas of the Seas around the British Isles*. Ministry of Agriculture, Fisheries and Food, London

Leopold, M F and Couperus, A S (1995) Sightings of Atlantic White-sided dolphins *Lagenorhynchus acutus* near the south-eastern limit of the known range in the North-East Atlantic. *Lutra*, **38**, 77-80

Leopold, M F, Wolf, P A and Van der Meer, J (1992) *Netherl. Journ. Sea Res.*, **29**, 395-402

Lewis, E J (1992) *Social cohesion and residency of bottle-nosed dolphins (Tursiops truncatus) in west Wales. An analysis of photographic data*. BSc Thesis, University of Leeds

Lewis, E J and Evans, P G H (1993) Comparative ecology of bottle-nosed dolphins *(Tursiops truncatus)* in Cardigan Bay and the Moray Firth. *European Research on Cetaceans*, **7**, 57-62

Lick, R and Piatkowski, U (1998) Stomach contents of a northern bottlenose whale *(Hyperoodon ampullatus)* stranded at Hiddensee, Baltic Sea. *J. Mar. Biol. Ass. U.K.*, **78**, 643-650

Lien, J and Barry, F (1990) Status of Sowerby's beaked whale, *Mesoplodon bidens*, in Canada. *Canad. Field-Natur.*, **104**, 125-130

Liret, C, Allali, P, Creton, P, Guinet, C and Ridoux, V (1994) Foraging activity pattern of bottlenose dolphins around Ile de Sein, France and its relationship with environmental parameters. *European Research on Cetaceans*, **8**, 188-191

Liret, C, Creton, P, Evans, P G H, Heimlich-Boran, J R and Ridoux, V (1998) *English and French coastal Tursiops from Cornwall to the Bay of Biscay, 1996*. Photo-Identification Catalogue. Project sponsored by Ministère de l'Environnement, France and Sea Watch Foundation, UK

Lockyer, C (1995) Investigations of aspects of the life history of the harbour porpoise, *Phocoena phocoena*, in British waters. *Rep. Int. Whal. Commn* (special issue16), 189-198

Martin, A R (1996) *Beluga Whales*. Colin Baxter Photography Ltd, Grantown-on-Spey

Martin, A R (in press) Narwhal, *Monodon monoceros*. In *New Handbook of British Mammals* (ed. S Harris). Academic Press, London

Martin, A R and Clarke, M R (1986) The diet of sperm whales *(Physeter macrocephalus)* captured between Iceland and Greenland. *J. Mar. Biol. Assoc. U.K.*, **66**, 779-790

Martin, A R and Rothery, P (1993) Reproductive parameters of female long-finned pilot whales *(Globicephala melas)* around the Faroe Islands. *Rep. Int. Whal. Commn* (special issue 14), 263-304

Martin, A R, Smith, T G and Cox, O P (1998) Dive form and function in belugas *(Delphinapterus leucas)* of the Canadian high arctic. *Polar Biology*, **20**, 218-228

Martins, H R, Clarke, M R, Reiner, F and Santos, R S (1985) A pygmy sperm whale, *Kogia breviceps* (Blainville, 1838) (Cetacea: Odontoceti) stranded on Faial Island, Azores, with notes on cephalopod beaks in stomach. *Arquipélago, Série Ciências da Natureza*, **6**, 63-70

Mate, B (1989) Watching habits and habitats from earth satellites. *Oceanus*, **32**, 14-18

Mate, B R and Stafford, M (1994) Movements and dive behavior of a satellite-monitored Atlantic white-sided dolphin *(Lagenorhynchus acutus)* in the Gulf of Maine. *Mar. Mamm. Sci.*, **10**, 116-121

Mayo, C A and Marx, M K (1990) Surface foraging behaviour of the North Atlantic right whale and associated plankton characteristics. *Can. J. Zool.*, **67**, 1411-1420

McAlpine, D F, Murison, L D and Hoberg, E P (1997) New records for the pygmy sperm whale, *Kogia breviceps* (Physeteridae) from Atlantic Canada with notes on diet and parasites. *Mar. Mamm. Sci.*, **13**, 701-704

McCann, C and Talbot, F H (1963) The occurrence of True's beaked whale (*Mesoplodon mirus* True) in South African waters, with a key to South African species of the genus. *Proc. Linn. Soc. Lond.*, **175**, 137-144

McMahon, T, Raine, R, Titov, O and Boychuk, S 1995. Some oceanographic features of north-eastern Atlantic waters west of Ireland. *ICES Journal of Marine Science*, **52**, 221-232

Mead, J G (1984) Survey of reproductive data for the beaked whales (Ziphiidae). *Report of the International*

Whaling Commission (special issue 6), 91-96

Mead, J G (1989a) Beaked whales of the genus *Mesoplodon*. In *Handbook of Marine Mammals. Vol. 4. River dolphins and the larger toothed whales* (eds S H Ridgway and R Harrison), pp. 349-430. Academic Press, London

Mead, J G (1989b) Bottlenose Whales *Hyperoodon ampullatus* (Forster, 1770) and *Hyperoodon planifrons* Flower, 1882. In *Handbook of Marine Mammals. Vol. 4. River dolphins and the larger toothed whales* (eds S H Ridgway and R Harrison), pp. 321-348. Academic Press, London

Mercer, M C (1975) Modified Leslie-DeLury population models of the long-finned pilot whale *(Globicephala melaena)* and annual production of the short-finned squid *(Illex illecebrosus)* based upon their interaction at Newfoundland. *J. Fish. Res. Bd Can.*, **32**, 1145-1154

Mikkelsen, A M H and Sheldrick, M (1992) The first recorded stranding of a melon-headed whale *(Peponocephala electra)* on the European coast. *J. Zool.*, Lond., **227**, 326-329

Montero, R and Martín, V (1992) First account on the biology of Cuvier's whales, *Ziphius cavirostris*, in the Canary Islands. *European Research on Cetaceans*, **6**, 97-99

NAMMCO (1993) North Atlantic Marine Mammal Commission Scientific Committee. *Report from the working group of bottlenose and killer whales.*

Nansen, F (1925) *Hunting and Adventure in the Arctic.* Duffield, New York

Nemoto, T (1964) School of baleen whales in the feeding areas. *Sci. Rep. Whales Res. Inst., Tokyo*, **18**, 89-110

Nordøy, E S, Folkow, L P, Mårtensson, P-E, and Blix, A S (1995) Food requirements of Northeast Atlantic minke whales. In *Whales, Seals, Fish and Man* (eds A S Blix, L Walløe and Ø Ulltang), pp. 307-317. Elsevier, Amsterdam

Nores, C and Pérez, M C (1982) Primera cita de un cachalote pigmeo, *Kogia breviceps* (Blainville, 1838) (Mammalia, Cetacea) para las costas peninsulares españolas. *Bol. Cien. Nat. I.D.E.A.*, no. 29, 307-317

Northridge, S, Tasker, M, Webb, A, Camphuysen, K and Leopold, M (1997) White-beaked *Lagenorhynchus albirostris* and Atlantic white-sided dolphin *L. acutus* distributions in Northwest European and U.S. North Atlantic waters. *Rep. Int. Whal. Commn*, **47**, 797-805

Northridge, S P, Tasker, M L, Webb, A and Williams, J M (1995) Seasonal distribution and relative abundance of harbour porpoises *Phocoena phocoena* (L.), white-beaked dolphins *Lagenorhynchus albirostris* (Gray) and minke whales *Balaenoptera acutorostrata* (Lacepède) in the waters around the British Isles. *ICES J. Mar. Sci.*, **52**, 55-66

North Sea Task Force (1993) *North Sea Quality Status Report 1993.* Oslo and Paris Commissions, International Council for the Exploration of the Sea, London

Odell, D K and McClune, K M (1999) False killer whale *Pseudorca crassidens* (Owen, 1846). In *Handbook of Marine Mammals, Vol. 6. The second book of dolphins and the porpoises.* (eds S H Ridgway and R J Harrison), pp. 213-243. Academic Press, London

Øien, N (1987) An observation of Risso's dolphins in Norwegian waters. *Fauna* (Oslo), **39**, 173

Øien, N (1990) Sighting surveys in the northeast Atlantic in July 1988: distribution and abundance of cetaceans. *Rep. Int. Whal. Commn*, **40**, 499-511

Osborne, R W (1986) A behavioral budget of Puget Sound killer whales. In *Behavioural biology of killer whales* (eds B C Kirkevold and J S Lockard), pp. 211-249. A R Liss, New York

Palka, D, Read, A and Potter, C (1997) Summary of current knowledge of White-sided Dolphins *(Lagenorhynchus acutus)* from U.S. and Canadian Atlantic waters. *Rep. Int. Whal. Commn*, **47**, 729-734

Papastavrou, V, Smith, S C and Whitehead, H (1989) Diving behavior of the sperm whale, *Physeter macrocephalus*, off the Galapagos Islands. *Can. J. Zool.*, **67**, 839-846

Pascoe, P L (1986) Size data and stomach contents of common dolphins, *Delphinus delphis*, near Plymouth. *J. Mar. Biol. Assoc. U.K.*, **66**, 319-322

Pattiaratchi, C, James, A and Collins, M (1986) Island wakes and headland eddies: a comparison between remotely sensed data and laboratory experiments. *Journal of Geophysical Research*, **92**, 783-794

Penas-Pantiño, X M and Piñeiro Seage, A (1989) *Cetáceos, Focas e Tartarugas Mariñas das Costas Ibéricas.* Da Sociedade Galega de Historia Natural, Santiago, Spain

Perrin, W F, Leatherwood, S and Collet, A (1994) Fraser's dolphin *Lagenodelphis hosei* Fraser, 1956. In *Handbook of Marine Mammals* (eds S H Ridgway and R Harrison), pp. 225-240. Academic Press, London

Perrin, W F (1982) Report of the workshop on identity, structure and vital rates of killer whale populations. *Rep. Int. Whal. Commn*, **32**, 617-631

Perrin, W F, Wilson, C E and Archer II, F I (1994) Striped dolphin *Stenella coeruleoalba* (Meyen, 1833). In *Handbook of Marine Mammals. Vol. 5. The first book of dolphins* (eds S H Ridgway and R Harrison), pp.

129-159. Academic Press, London

Perryman, W L, Au, D W K, Leatherwood, S and Jefferson, T A (1994) Melon-headed whale *Peponocephala electra* Gray, 1846. In *Handbook of Marine Mammals. Vol. 5. The first book of dolphins* (eds S H Ridgway and R Harrison), pp. 363-386. Academic Press, London

Pihl, S and Frikke, J (1992) Counting birds from aeroplane. In *Manual for aeroplane and ship surveys of waterfowl and seabirds* (eds J Komdeur, J Bertelsen and G Cracknell), pp. 8-23. IWRB Special Publication No. 19, Slimbridge

Pingree, R D and Griffiths, D K (1978) Tidal fronts on the shelf seas around the British Isles. *Journal of Geophysical Research*, **83**, 4615-4622

Pingree, R D and Mardell, G T (1986) Coastal tidal jets and tidal fringe development around the Isles of Scilly. *Estuarine, Coastal and Shelf Science*, **23**, 581-594

Pingree, R D, Holligan, P M and Mardell, G T (1978) The effects of vertical stability on phytoplankton distribution in the summer on the northwest European Shelf. *Deep-Sea Research*, **25**, 1011-1028

Plön, S E E, Bernard, R T F, Klages, N T K and Cockcroft, V G (1999) Stomach content analysis of pygmy and dwarf sperm whales and its ecological implications: is there niche partitioning? *European Research on Cetaceans*, **13**, 336-339

Pollock, C, Reid, J B, Webb, A and Tasker, M (1997) The distribution of seabirds and cetaceans in the waters around Ireland. *JNCC Report*, No. 267. Joint Nature Conservation Committee, Aberdeen

Pollock, C M, Mavor, R, Weir, C R, Reid, A, White, R W, Tasker, M L, Webb, A and Reid, J B (2000) *The distribution of seabirds and marine mammals in the Atlantic Frontier, north and west of Scotland.* Joint Nature Conservation Committee, Aberdeen

Rae, B B (1973) Additional notes on the food of the common porpoise *(Phocoena phocoena)*. *J. Zool.*, Lond., **169**, 127-131

Raine, R, O'Mahony, J, McMahon, T and Roden, C (1990) Hydrography and phytoplankton of waters off south-west Ireland. *Estuarine, Coastal and Shelf Science*, **30**, 579-592

Read, A J (1999) Harbour porpoise *Phocoena phocoena* (Linnaeus, 158). In *Handbook of marine mammals. Vol. 6. The second book of dolphins and the porpoises* (eds S H Ridgway and R Harrison), pp. 323-356. Academic Press, London

Reeves, R R, Dietz, R and Born, E W (1994) Overview of the special issue 'Studies of white whales *(Delphinapterus leucas)* and narwhals *(Monodon monoceros)* in Greenland and adjacent waters'. *Meddelelser om Grønland, Bioscience* **39**, 3-11

Reeves, R R and Mitchell, E D (1987) Distribution, migration, exploitation and former abundance of white whales in Baffin Bay and adjacent waters. *Canad. Spec. Publ. of Fish. and Aquat. Sci.*, **99**. 34 pp.

Reeves, R R, Mitchell, E and Whitehead, H (1993) Status of the Northern Bottlenose whale, *Hyperoodon ampullatus. Canad. Field-Natur.*, **107**, 490-508

Reeves, R R, Smeenk, C, Brownell, Jr, R L and Kinze, C C (1999a) Atlantic white-sided dolphin *Lagenorhynchus acutus* Gray, 1828. In *Handbook of marine mammals. Vol. 6. The second book of dolphins and the porpoises* (eds S H Ridgway and R Harrison), pp. 31-56. Academic Press, London

Reeves, R R, Smeenk, C, Kinze, C C, Brownell, Jr, R L and Lien, J (1999b) White-beaked dolphin *Lagenorhynchus albirostris* Gray, 1846. In *Handbook of Marine Mammals. Vol. 6. The second book of dolphins and the porpoises* (eds S H Ridgway and R Harrison), pp. 1- 30. Academic Press, London

Reiner, F (1981) Nota sobre a ocorréncia de um cachalote-anão *Kogia breviceps* Blainville na Praia de Salgueiros - Vila Nova de Gaia. *Memórias do Museo do Mar, série zoológica*, **2**(15), 1-12

Reiner, F (1986) First record of Sowerby's beaked whale from the Azores. *Sci. Rep. Whales Res. Inst.*, Tokyo, **37**, 103-107

Reiner, F, Gonçalves, J M and Santos, R S (1993) Two new records of Ziphiidae (Cetacea) for the Azores with an updated checklist of cetacean species. *Arquipélago* (Life and Marine Sciences), **11A**, 113-118

Reiner, F M E, Dos Santos, M E and Wenzel, F W (1996) Cetaceans of the Cape Verde archipelago. *Mar. Mamm. Sci.*, **12**, 434-443

Relini, L O, Cappello, M and Poggi, R (1994a) The stomach content of some bottlenose dolphins *(Tursiops truncatus)* from the Ligurian Sea. *European Research on Cetaceans*, **8**, 192-195

Relini, L O, Relini, G, Cima, C, Palandri, G, Relini, M and Torchia, G (1994b) *Meganyctiphanes norvegica* and fin whales in the Ligurian Sea: new seasonal pattern. *European Research on Cetaceans*, **8**, 179-182

Relini, L O, Relini, G, Palandri, G, Relini, M, Garibaldi, F, Cima, C, Torchia, G and Costa, C (1998) Notes on ecology of the Mediterranean krill, a mirror of the behaviour of Mediterranean fin whales. *European Research on Cetaceans*, **12**, 119

Rice, D W (1989) Sperm Whale *Physeter macrocephalus* Linnaeus, 1758. In *Handbook of Marine Mammals.*

Volume 4. River dolphins and the larger toothed whales (eds S H Ridgway and R Harrison), pp. 177-233. Academic Press, San Diego, CA

Ritter, F and Brederlau, B (1999) Behavioural observations of dense-beaked whales *(Mesoplodon densirostris)* off La Gomera, Canary Islands (1995-1997). *Aquatic Mammals*, **25**, 55-61

Robinson, B H and Craddock, J E (1983) Mesopelagic fishes eaten by Fraser's dolphin, *Lagenodelphis hosei. Fish. Bull., U.S.*, **81**, 283-289

Rogan, E, Baker, J R, Jepson, P D, Berrow, S and Kiely, O (1997) A mass stranding of white-sided dolphins *(Lagenorhynchus acutus)* in Ireland: biological and pathological studies. *J. Zool.*, Lond., **242**, 217-227

Rosen, M R, Evans, P G H, Boran, J R, Bell, G and Thomas, C (2000) Cetacean studies in the Celtic Sea, English Channel and S.W. North Sea: using training surveys for data collection. *European Research on Cetaceans*, **14**, 383-386

Ross, G J B (1969) Evidence for a southern breeding population of True's beaked whale. *Nature*, London, **222**, 585

Ross, G J B (1984) The smaller cetaceans of the south east coast of southern Africa. *Annals of the Cape Provincial Museums of Natural History*, **15**, 173-410

Ross, H M and Wilson, B (1996) Violent interactions between bottlenose dolphins and harbour porpoises. *Proc. R. Soc. Lond.* B, **263**, 283-286

Salomón, O, Blanco, C and Raga, J A (1997) Diet of the bottlenose dolphin *(Tursiops truncatus)* in the Gulf of Valencia (Western Mediterranean). *European Research on Cetaceans*, **11**, 156

Sanders-Reed, C A, Hammond, P S, Grellier, K and Thompson, P M (1999) Development of a population model for bottlenose dolphins. *Scottish Natural Heritage Research, Survey and Monitoring Report*, No. 156, 1-38, Edinburgh

Santos Vázquez, M B (1998) *Feeding ecology of harbour porpoises, common and bottlenose dolphins and sperm whales in the Northeast Atlantic*. PhD thesis, University of Aberdeen

Santos Vázquez, M B, Pierce, G J, Boyle, P R, Reid, R J, Ross, H M, Patterson, I A P, Kinze, C C, Tougaard, S, Lick, R, Piatkowski, U and Hernández García, V (1999) Stomach contents of sperm whales *Physeter macrocephalus* stranded in the North Sea 1990-1996. *Mar. Ecol. Progr. Series.*, **183**, 281-294

Santos Vázquez, M B, Pierce, G J, Herman, J, López, A, Guerra, A, Mente, E and Clarke, M R (2001) Feeding ecology of Cuvier's beaked whale *(Ziphius cavirostris)*: a review with new information on the diet of this species. *J. Mar. Biol. Ass. U.K.*, **81**, 687-694

Santos Vázquez, M B, Pierce, G J, Ross, H M and Reid, R J (1994) Diets of small cetaceans from the Scottish coast. ICES Marine Mammal Committee C.M. 1994/No. 11, 1-16

Sarvas, T H and Fleming, V (1999) The effects of the deep scattering layer on the diving behaviour of sperm whales off Andøya, Norway. *European Research on Cetaceans*, **13**, 341-345

Schweder, T, Skaug, H J, Dimakos, X K, Langaas, M and Øien, N (1997) Abundance of Northeastern Atlantic minke whales, estimates for 1989 and 1995. Rep. *Int. Whal. Commn*, **47**, 453-483

Seaman, G A, Lowry, L F and Frost, K J (1982) Foods of belukha whales *(Delphinapterus leucas)* in western Alaska. *Cetology*, **44**, 1-19

Sears, R (1983) A glimpse of blue whale feeding in the Gulf of St. Lawrence. *Whalewatcher*, **17**, 12-14

Sears, R, Williamson, J M, Wenzel, F W, Bérubé, M, Gendron, D and Jones, P(1990) Photographic identification of the blue whale *(Baaenoptera musculus)* in the Gulf of St. Lawrence, Canada. *Rep. Int. Whal. Commn* (special issue 12), 335-342

Selzer, L A and Payne, P M (1988) The distribution of white-sided *(Lagenorhynchus acutus)* and common dolphins *(Delphinus delphis)* vs. environmental features of the continental shelf of the northeastern United States. *Mar. Mamm. Sci.*, **4**, 141-153

Sequeira, M, Inácio, A, Silva, M A and Reiner, F (1992) Arrojamentos de Mamíferos Marinhos na Costa Portuguesa entre 1978 e 1988. *Estudos de Biologia e Conversação da Natureza*, **7**, 1-48

Sequeira, M, Inácio, A, Silva, M A and Reiner, F (1996) *Arrojamentos de Mamíferos Marinhos na Costa Continental Portuguesa entre 1989 e 1994*. Estudos de Biologia e Conversação da Natureza, No. 19

Sequiera, M (1996) Harbour porpoises, *Phocoena phocoena*, in Portuguese waters. *Rep. Int. Whal. Commn.*, **46**, 583-586

Sergeant, D E (1982) Mass strandings of toothed whales (Odontoceti) as 'a population phenomenon'. *Sci. Rep. Whales Res. Inst., Tokyo*, **34**, 1-47

Sergeant, D E, St Aubin, D J, and Geraci, J R (1980) Life history and northwest Atlantic status of the Atlantic white-sided dolphin, *Lagenorhynchus acutus. Cetology*, **37**, 1-12

Sheldrick, M (1989) Stranded whale records for the entire British coastline, 1967-1986. *Invest. on Cetacea* (ed. G. Pilleri), **22**, 298-329

Sigurjónsson, J (1992) Recent studies on abundance and trends in whale stocks in Icelandic and adjacent waters. *Proc. Roy. Acad. Overs. Sci.* (Brussels), 1992, 77-111

Sigurjónsson, J (1995) On the life history and autecology of North Atlantic rorquals. In *Whales, Seals, Fish and Man* (eds A S Blix, L Walløe and Ø Ulltang), pp. 425-441. Elsevier Science, New York

Sigurjónsson, J and Gunnlaugsson, T (1990) Recent trends in abundance of blue *(Balaenoptera musculus)* and humpback whales *(Megaptera novaeangliae)* of west and southwest Iceland, with a note on occurrence of other cetacean species. *Rep. Int. Whal. Commn*, **40**, 537-551

Silva, M A (1999) Diet of common dolphin *Delphinus delphis*, off the Portuguese continental shelf. *J. Mar. Biol. Assoc. U.K.*, **79**, 531-540

Silva, M A and Sequeira, M (1997) Stomach contents of marine mammals stranded on the Portuguese coast. *European Research on Cetaceans*, **11**, 176-179

Similä, T (1997) *Behavioural Ecology of Killer Whales in Northern Norway*. PhD Thesis, Norwegian College of Fisheries Science, University of Tromsø, Norway

Simpson, J H (1981) The shelf-sea fronts: implications of their existence and behaviour. *Philosophical Transactions of the Royal Society of London, A*, **302**, 531-546

Smeenk, C, Leopold, M F and Addink, M J (1992) Note on the harbour porpoise *Phocoena phocoena* in Mauretania, West Africa. *Lutra*, **35**, 98-104

Smith, T D, Allen, J, Clapham, P J, Hammond, P S, Katona, S, Larsen, F, Lien, J, Mattila, D, Palsbøll, P J, Sigurjónsson, J, Stevick, P T and Øien, N (1999) An ocean-basin wide mark-recapture study of the North Atlantic humpback whale *(Megaptera novaeangliae)*. *Mar. Mamm. Sci.*, **15**, 1-32

Smolker, R A, Richards, A F, Connor, R C and Pepper, J W (1992) Sex differences in patterns of association among Indian Ocean bottlenose dolphins. *Behaviour*, **123**, 38-69

Steiner, L, Gordon, J and Beer, C J (1998) Marine mammals of the Azores. *European Research on Cetaceans*, **12**, 79

Stone, C J (1997) Cetacean observations during seismic surveys in 1996. *JNCC Report, No. 228*. Joint Nature Conservation Committee, Peterborough

Stone, G S, Katona, S K, Mainwaring, A, Allen, J M and Corbett, H D (1992) Respiration and surfacing rates of fin whales, *Balaenoptera physalus*, observed from a lighthouse tower. *Rep. Int. Whal. Commn*, **42**, 739-745

Tasker, M L, Jones, P H, Dixon, T J and Blake, B F (1984) Counting seabirds at sea from ships: A review of methods employed and a suggestion for a standardized approach. *Auk*, **101**, 567-577

Thompson, d'A W (1928) On whales landed at the Scottish whaling stations during the years 1908-1914 and 1920-1927. *Scientific Investigations, Fishery Board of Scotland*, **3**, 1-40

Tiedemann, R (1996) Mitochondrial DNA sequence patterns of harbour porpoises *(Phocoena phocoena)* from the North and the Baltic Sea. *Zeitschr. für Säugetierkunde*, **61**, 104-111

Tregenza, N J C, Berrow, S D, Hammond, P S and Leaper, R (1997) Harbour porpoise (*Phocoena phocoena* L.) by-catch in set gillnets in the Celtic Sea. *ICES Journal of Marine Science*, **54**, 896-904

Turrell, W R, Henderson, E W, Slesser, G, Payne, R and Adams, R D (1992) Seasonal changes in the circulation of the northern North Sea. *Continental Shelf Research*, **12**, 257-286

Ugarte, F and Similä, T (1993) Surface and underwater observations of co-operatively feeding killer whales in northern Norway. *Can. J. Zool.*, **71**, 1494-1499

Valverde, J A and Galán, J M (1996) Notes on a specimen of Gervais' beaked whale *Mesoplodon europaeus* (Gervais), Ziphiidae, stranded in Andalucia, southern Spain. *European Research on Cetaceans*, **10**, 177-183

Van Bree, P J H, Collet, A, Desportes, G, Hussenot, E and Raga, J A (1986) Le dauphin de Fraser *Lageodelphis hosei* (Cetacea, Odontoceti), espèce nouvelle pour la faune d'Europe. *Mammalia*, **50**, 57-86

Van Oort, E D (1926) Over eenige aan de kust van Nederland waargenomen cetaceeën-soorten. *Zool. Meded.*, Leiden, **9**, 211-213

Van Utrecht, W L (1978) Age and growth in *Phocoena phocoena* Linnaeus, 1758 (Cetacea, Odontoceti) from the North Sea. *Bijdragen tot de Dierkunde*, **48**, 16-28

Víkingsson, G (1993) Northern Bottlenose whale *(Hyperoodon ampullatus)*. Availability of data and status of research in Iceland. NAMMCO SC-WG/NBK1/7 1-4

Vinther, M (1999). Bycatches of harbour porpoises (*Phocoena phocoena* L.) in Danish set net fisheries. *J. Cetacean Res. Manage.*, **1**, 123-136

Vonk, R and Martin, M V (1989) Goose-beaked whales *Ziphius cavirostris* mass strandings in the Canary Islands. *European Research on Cetaceans*, **3**, 73-77

Walker, D, Diamond, J and Stokes, J (2001) True's beaked whale - a first confirmed live sighting for the eastern North Atlantic. *Orca*, **2**, 63-66

Walton, M J (1997) Population structure of harbour porpoises *Phocoena phocoena* in the seas around the U.K. and adjacent waters. *Proc. Roy. Soc. Lond., Series B*, **264**, 89-94

Waring, G T, Palka, D L, Clapham, P J, Swartz, S, Rossman, M C, Cole, T V N, Bisack, K D and Hansen, L J (1999) U.S. Atlantic Marine Mammal Stock Assessments - 1998. *NOAA Technical Memorandum NMFS-NE-116*. National Marine Fisheries Service, Woods Hole

Waring, G T, Gerrior, P, Payne, P M, Parry, B L and Nicholas, J R (1990) Incidental take of marine mammals in foreign fishery activities off the northeast United States, 1977-1988. *Fish. Bull. U.S.*, **88**, 347-360

Waters, S and Whitehead, H (1990) Aerial behavior in sperm whales. *Can. J. Zool.*, **68**, 2076-2082

Watkins, W A, Daher, M A, Fristrup, K M, Howald, T J and Notarbartolo di Sciara, G (1993) Sperm whales tagged with transponders and tracked underwater by sonar. *Mar. Mamm. Sci.*, **9**, 55-67

Watkins, W A, Moore, K E and Tyack, P (1985) Sperm whale acoustic behaviors in the southeast Caribbean. *Cetology*, **49**, 1-15

Webb, A and Durinck, J (1992) Counting birds from ship. In *Manual for aeroplane and ship surveys of waterfowl and seabirds* (eds J Komdeur, J Bertelsen and G Cracknell), pp 24-37. IWRB Special Publication No. 19, Slimbridge

Webb, A, Harrison, N M, Leaper, G M, Steele, R D, Tasker, M L and Pienkowski, M W (1990) *Seabird distribution west of Britain*. Final Report of Phase 3 of the Nature Conservancy Council Seabirds at Sea Project November 1986-March 1990. Nature Conservancy Council, Peterborough

Wells, R S and Scott, M D (1999) Bottlenose dolphin *Tursiops truncatus* (Montagu, 1821). In *Handbook of marine mammals. Vol. 6. The second book of dolphins and the porpoises* (eds S H Ridgway and R Harrison), pp. 137-182. Academic Press, London

Wells, R S, Scott, M D and Irvine, A B (1987) The social structure of free-ranging bottlenose dolphins. In *Current Mammalogy* (ed. H Genoways), pp. 263-317. Plenum Press, New York

White, R W, Gillon, K W, Black, A D and Reid, J B (2002) *The distribution of seabirds and marine mammals in Falkland Islands waters*. Joint Nature Conservation Committee, Peterborough

Whitehead, H (1987) Social organization of sperm whales off the Galapagos: implications for management and conservation. *Rep. Int. Whal. Commn*, **37**, 195-199

Whitehead, H and Carlson, C (1988) Social behavior of feeding finback whales off Newfoundland: comparisons with the sympatric humpback whale. *Can. J. Zool.*, **66**, 217-221

Whitehead, H and Moore, M J (1982) Distribution and movements of West Indian humpback whales in winter. *Can. J. Zool.*, **60**, 2203-2211

Whitehead, H , Gowans, S, Faucher, A and McCarrey, S W (1997) Population analysis of northern bottlenose whales in the Gully, Nova Scotia. *Mar. Mamm. Sci.*, **13**, 173-185

Whitehead, H and Weilgart, L (2000) The Sperm Whale. In *Cetacean Societies* (eds J Mann, R C Connor, P L Tyack and H Whitehead), pp. 154-172. University of Chicago Press, Chicago/London

Williams, A D, Williams, R, Heimlich-Boran, J R, Evans, P G H, Tregenza, N J C, Ridoux, V, Liret, C and Savage, S (1996) A preliminary report on an investigation into bottlenose dolphins *(Tursiops truncatus)* of the English Channel: A collaborative approach. *European Research on Cetaceans*, **10**, 217-220

Wilson, B, Thompson, P M and Hammond, P S (1997) Habitat use by bottlenose dolphins: seasonal distribution and stratified movement patterns in the Moray Firth, Scotland. *J. Appl. Ecol.*, **34**, 1365-1374

Winn, H E and Reichley, N E (1985) Humpback whale *Megaptera novaeangliae*. In *Handbook of Marine Mammals. Volume 3. The Sirenians and Baleen Whales* (eds S H Ridgway and R Harrison), pp. 241-274. Academic Press, London

Wood, C (1998) Movement of bottlenose dolphins around the south-west of Britain. *J. Zool.*, Lond., **246**, 155-163

Woodley, T H and Gaskin, D E (1996) Environmental characteristics of North Atlantic right and fin whale habitat in lower Bay of Fundy, Canada. *Can. J. Zool.*, **74**, 75-84

Yochem, P and Leatherwood, S (1985) The blue whale. In *Handbook of Marine Mammals. Volume 3. The Sirenians and Baleen Whales*. (eds S H Ridgway and R Harrison), pp. 193-240. Academic Press, London

Young, K, Freitas, L and Mclanaghan, R (1999) Sowerby's beaked whale *(Mesoplodon bidens)* sighting off the Island of Madeira. *European Research on Cetaceans*, **13**, 281

Zachariassen, P (1993) Pilot whale catches in the Faroe Islands, 1709-1992. *Rep. Int. Whal. Commn* (special issue 14), 69-88

Zonfrillo, B, Sutcliffe, R, Furness, R W and Thompson, D R (1988) Notes on a Risso's dolphin from Argyll, with analyses of its stomach contents and mercury levels. *Glasgow Naturalist*, 1988, 297-303